Phase-Locked Loops
Principles and Practice

Phase-Locked Loops

Principles and Practice

Paul V. Brennan

Department of Electronic and Electrical Engineering
University College London

McGraw-Hill

New York   San Francisco   Washington, D.C.   Auckland   Bogotá
Caracas   Lisbon   London   Madrid   Mexico City   Milan
Montreal   New Delhi   San Juan   Singapore
Sydney   Tokyo   Toronto

McGraw-Hill

A Division of The **McGraw·Hill** Companies

1 2 3 4 5 6 7 8 9 0   DOC/DOC   9 0 1 0 9 8 7 6

ISBN 0-07-007568-9

First published 1996 by MACMILLAN PRESS LTD
Houndmills, Basingstoke, Hampshire RG21 2XS
and London.

Printed and bound by R. R. Donnelley & Sons Company.

For my parents and
for my wife, Francine

Contents

Preface

This book is intended as a concise guide to the theory and design of phase-locked loop (PLL) circuits. It is written from an engineering point of view, including numerous block diagrams, example circuits and experimental results, whilst making use of 'engineering' analytical methods such as signal flow graphs and Laplace transforms to retain a firm grip on the theoretical basis. Competent PLL design and analysis requires a considerable degree of theoretical understanding and yet it is a subject which is seldom formally taught to any depth. For this reason the book has a strong theoretical foundation, with most results derived from first principles, although mathematics is included for reasons of practical relevance rather than academic interest. This, hopefully, has resulted in a substantially self-contained text which should prove useful to both those new to and to those more familiar with the subject.

The initial motivation for the work came from a lecture course which I have been giving on PLLs since 1990 in the Department of Electronic and Electrical Engineering at University College London. This may explain the tutorial nature of parts of the book (such as the examples and worked solutions) which is slightly unusual in a PLL book and which should appeal to third/fourth year undergraduate and postgraduate students.

Although the book is relatively brief, a number of quite advanced analyses and sophisticated techniques are included, such as fractional-N synthesis. I have endeavoured to treat the subject in a straightforward yet comprehensive fashion with the overall aim of providing a coherent, consistent reference which avoids cutting too many corners. A spin-off of this approach is that there are one or two techniques and treatments which I believe to be original to this text.

I would like to thank Brian Oughton for first introducing me to PLLs in 1984 and those who have since given me the opportunity of continuing my interest in this fascinating, elegant and very useful area of electronic circuits.

Paul V. Brennan

Phase-Locked Loops
Principles and Practice

1 Introduction

Phase-locked loops (PLLs) are a well established and very widely used circuit technique in modern electronic systems. They involve a curious blend of analogue and digital techniques, operate with both linear and non-linear behaviour and are amenable to continuous, sampled or even chaotic analysis. They also result in amongst the most elegant of electronic circuit designs. Although occasionally regarded as something of a 'black art', the truth is that the subject is most amenable to accurate analysis and rational design once the fundamental principles are clearly understood.

In essence, PLLs are circuits in which the phase of a local oscillator is maintained close (or locked) to the phase of an external signal. The technique was first developed in the 1930s as a means of implementing a zero IF frequency synchronous receiver – a device which has never actually achieved widespread use. Examples of the many successful applications of PLLs include line synchronisation and colour sub-carrier recovery in TV receivers, local oscillators and FM or PM demodulators in radio receivers and frequency synthesisers in transceivers (such as mobile telephones) and signal generators, to name but a few. The basic operation of a PLL is deceptively simple, however the detailed design of a PLL circuit for a particular application often requires a great deal of understanding of the underlying principles of operation, circuit subtleties and associated limitations. It is very easy to design a PLL badly by ignoring basic control loop considerations or by attempting to make short-cuts. The purpose of this book is to introduce the fundamental principles of PLLs, the basic design considerations and some of their many properties and applications. As far as possible, ideas and concepts are built up from first principles, to provide a full understanding of the nature of this very useful, often intriguing and versatile electronic building block.

A PLL is a control loop consisting of four fundamental components, as shown in figure 1.1. These are a phase detector (sometimes, and perhaps more precisely, known as a phase comparator), loop filter, voltage controlled oscillator (VCO) and a frequency divider. These components are connected in a simple feedback arrangement in which the phase detector compares the phase, ϕ_i, of an input signal with the phase, ϕ_o/N, of the feedback VCO signal. The phase detector output voltage is dependent on the difference in phase of the two applied inputs and is used to adjust the VCO until this phase difference is very small. Ideally, the phase detector and VCO responses are linear but, in practice, the phase detector response is cyclic and the VCO response is not perfectly linear. The loop is then in a stable equilibrium so that the VCO phase is locked to the input signal phase, $\phi_o = N\phi_i$. Thus the circuit behaves as a phase multiplier and also, since frequency is the time derivative of phase, as a frequency multiplier. By varying the division ratio, the output frequency may be stepped in multiples of the input frequency – giving a basic frequency synthesiser. Alternatively, by introducing signals into the loop, phase or frequency modulation around a stable carrier frequency may be achieved. It soon becomes apparent that the same basic arrangement of figure 1.1 lends itself to a wide variety of applications, often involving considerable subtlety of design.

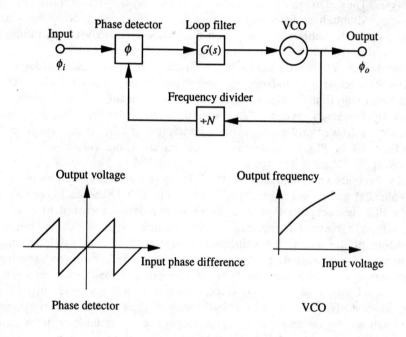

Figure 1.1 Basic arrangement of a phase-locked loop

The behaviour of a PLL can be likened to that of a feedback voltage amplifier such as the simple example shown in figure 1.2. This is the block diagram representation of a non-inverting op-amp voltage amplifier, where the output and input voltage are related by $V_o = KV_i$. It is similar to the basic PLL except that it uses a voltage comparator instead of a phase comparator, a potential divider in the feedback instead of a frequency divider and the controlled parameter is voltage rather than phase. The voltage comparator provides an error signal dependent on the difference between the output voltage divided by K and the input voltage in a similar manner to the operation of the phase comparator in the equivalent PLL circuit of figure 1.1. The analogy between these two circuits has been drawn to illustrate that PLLs are simply a type of control loop in which the parameter under control happens to be the phase of a number of signals. As such, PLLs may be designed and analysed with standard control methods, although it must be said that some of the loop components, most notably the phase detector, do have unusual properties!

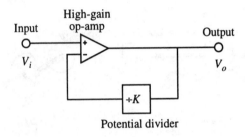

Figure 1.2 Voltage amplifier based on an op-amp

A simple illustration of a PLL circuit is given in figure 1.3. This represents the block diagram of a motor speed controller in which the speed of a motor, f_o revs/s, is precisely locked to an accurate reference frequency, f_{ref}, obtained from a crystal oscillator. One application of such a technique is as a turntable speed controller in audio electronics. In order to control the motor speed it is first necessary, of course, to measure the speed. This may be achieved with the aid of a tachometer to generate a pulsed signal representing the instantaneous angular velocity of the motor. Angular velocity in this example is analogous to frequency in an oscillator. The tachometer may comprise a magnetically encoded disc attached to the drive shaft in close proximity to a magnetic pick-up coil (similar to a tape head) or an optical disc with radially alternating transparent and opaque areas coupled to a light source and photodiode. The tachometer output is subsequently amplified, applied to a frequency divider and then presented to a phase detector which compares the angular speed of the motor shaft with the frequency of a stable source derived from a crystal oscillator. The phase detector output is dependent on the phase difference between its two inputs and is applied to a filter with integrator-type properties which eventually amplifies any small DC signal from the phase detector to a large voltage necessary to provide the correct motor speed. Adjustment of motor speed may be achieved by varying the field current, supply voltage or supply frequency. If the shaft encoder produces K pulses per revolution and the frequency divider ratio is N then the frequencies of the two signals at the phase detector are equal when $f_{ref} = Kf_o/N$ and this is the condition for stable operation of the system. If a variable ratio divider is used, then the motor speed may be varied in multiples of f_{ref}/K. As an example, the values in figure 1.3 of $f_{ref} = 1$ kHz, $K = 1000$ and $N = 10 - 100$ illustrate a design capable of producing motor speeds varying over a useful range from 600 to 6000 rpm in steps of 60 rpm.

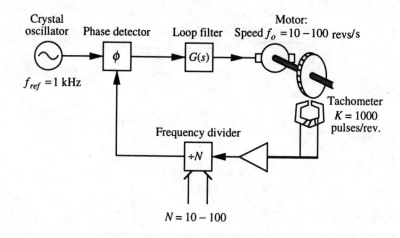

Figure 1.3 A simple motor speed controller

Whilst the simple example of figure 1.3 demonstrates the principles of PLLs in a static, settled situation it overlooks dynamic aspects of behaviour – such as transient response and modulation capability. Such active characteristics of PLLs are governed by the loop filter properties which is a key area of PLL design and is responsible for a whole range of loop specifications which almost invariably must be traded-off against each other in any given application. For instance, a narrow loop bandwidth is needed in applications where the input signal is noisy – such as with FM discriminators or carrier-recovery circuits, however this tends to result in slow loop operation which may impose an unacceptable limit on the frequency response or tuning time. A principal attraction of PLLs is that the loop bandwidth may be made very narrow so that the loop can recover signals buried in noise and yet still track their frequency variations over a wide range. On the other hand, a wider bandwidth is needed in applications requiring higher operating speed, such as agile frequency synthesisers and wide-band modulators. There are a number of fundamental limitations and also certain practical constraints on loop performance which are equally important when considering PLL design. These are amongst the points which are considered in the following chapters.

2 Loop Components

Before embarking on an account of the overall loop operation, the four main loop components of figure 1.1 will now be described in some detail. Further information on these and other more exotic loop components may be found in a number of other sources including references 1 to 3.

2.1 Phase detectors

Phase detectors are arguably the most interesting and important part of any PLL. They are a form of comparator providing a DC output signal proportional to the difference in phase between two input signals. This may be written,

$$V_p = K_p(\phi_{i1} - \phi_{i2}) \tag{2.1}$$

where V_p is the output voltage, ϕ_{i1} and ϕ_{i2} are the phases of the input signals and K_p is the phase detector gain in volts per radian. Although a linear response would be ideal, in practice the response of phase detectors is non-linear and repeats in a cyclic fashion over a limited phase range. However, the response is usually very nearly linear in a narrow phase range close to the point at which the loop would normally lock, and the slope of the phase detector characteristic, the *phase detector gain*, is of most interest at this point.

There are two basic types of phase detector, *multiplier* and *sequential*. Multiplier phase detectors, as the name implies, form the product of two alternating input signals, the DC component of which is dependent on the phase difference between the inputs. Sequential phase detectors, on the other hand, respond to the relative timing of the edges of the input signals and are therefore implemented in digital form. The multiplier type, being linear, are useful in applications where the input signal is noisy, since the output S/N ratio degrades at the same rate as the input S/N ratio. The sequential type are vulnerable to poor operation under low S/N conditions in which they exhibit a threshold effect and so they are normally used at S/N ratios significantly above 10 dB. However, they can offer far superior capture and tracking performance.

2.2 Multiplier phase detectors

The simplest example of a multiplier type phase detector, which is actually very widely used, is the analogue multiplier of figure 2.1. This is basically a double-balanced mixer (or four quadrant multiplier) with a DC coupled output port, that is used to produce the product of two input signals $V_1(t)$ and $V_2(t)$. Of course, a standard double-balanced mixer (with a DC coupled IF port) could be used as such a phase detector although mixers are produced especially for this purpose with low DC offsets and high sensitivity – both of which are important characteristics of multiplier type phase detectors.

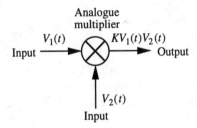

Figure 2.1 Analogue multiplier phase detector

Operation can be easily understood by considering sinusoidal input signals, $V_1(t) = V_1\cos\omega t$ and $V_2(t) = V_2\cos(\omega t - \phi)$, the phase detector output being proportional to the product of these signals,

$$V_p(t) = KV_1(t)V_2(t) = \frac{KV_1V_2}{2}\left[\cos(2\omega t - \phi) + \cos\phi\right]$$

Thus the output consists of a DC term and a double-frequency component. The double-frequency component is filtered out by the action of the loop filter and is of no significance, leaving

$$V_p = \frac{KV_1V_2}{2}\cos\phi \qquad (2.2)$$

This shows that the phase detector output varies sinusoidally with phase difference, with zeros at $\phi = \pi/2 + n\pi$. Assuming that the loop is based on a high-gain filter (such as a form of integrator) then it will lock when the phase detector output is zero. The loop therefore locks, with this particular type of phase detector, when there is a quadrature phase difference between the phase detector inputs. Phase detector gain is given by

$$\frac{dV_p}{d\phi} = \frac{d}{d\phi}\left[\frac{KV_1V_2}{2}\cos\phi\right] = \frac{KV_1V_2}{2}\sin\phi$$

and in the linear regions of the characteristic, around $\phi = \pi/2 + n\pi$, the gain in V/rad is

$$K_p = \frac{KV_1V_2}{2} \tag{2.3}$$

So the phase detector gain is equal to the peak value of output voltage. It is important to know this quantity when designing the loop filter (section 2.7) since it represents one of the scaling factors in the loop gain. The usable range of the detector, where operation is approximately linear (or at least monotonic), is limited to within $\pm\pi/2$ rads of $\phi = \pi/2 + n\pi$.

Analogue multiplier type phase detectors are often used with one square-wave input signal and one sinusoidal input signal. This is often for reasons of convenience since the frequency divider and maybe also VCO are implemented in digital circuitry. In this case, for relatively low frequency operation (up to a few MHz), the circuit of figure 2.2 is useful. It performs exactly the same function as an analogue multiplier driven by one analogue signal and one digital signal, but has the advantage that it requires no analogue multiplying components.

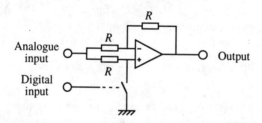

Figure 2.2 Multiplying phase detector with analogue and digital inputs

The circuit works simply by acting as a unity gain inverting amplifier when the digital input is high and a unity gain non-inverting amplifier when the digital input is low. The performance of this mode of operation of analogue multiplier phase detectors is best explained with reference to the timing diagrams of figure 2.3. Here the inputs of an analogue multiplier phase detector are shown as a sinewave of amplitude V_1 and a square-wave, lagging by ϕ, of amplitude V_2. Now, when the square-wave input is high the sinusoid is non-inverted and when it is low the sinusoid is inverted. Thus the phase detector output is a double-frequency sample of a sinusoid, $\cos\theta$, over the range $\theta = \phi - \pi/2$ to $\theta = \phi + \pi/2$. This signal has an obvious DC component which is dependent on the phase difference, ϕ, as follows

$$V_p = \overline{KV_1(t)V_2(t)} = \frac{KV_1V_2}{\pi} \int\limits_{\phi-\pi/2}^{\phi+\pi/2} \cos\theta \; d\theta = \frac{2KV_1V_2}{\pi}\cos\phi \qquad (2.4)$$

from which it is clear that the response is sinusoidal, as before, but with a phase detector gain of $K_p = 2KV_1V_2/\pi$ V/rad. At this point it should be noted that successive half-cycles of the output waveform of figure 2.3 are identical only if a square-wave of 50% duty cycle is used – and this is the optimum duty cycle for this type of phase detector.

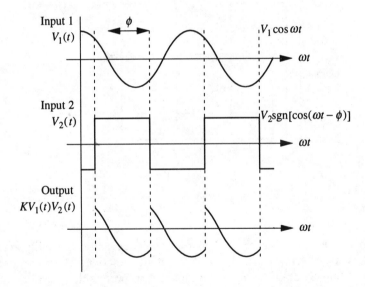

Figure 2.3 Analogue multiplier phase detector operation with mixed sinusoidal and square-wave inputs

The phase detector may also be used with square-wave signals applied to both inputs, as shown in figure 2.4. The output signal is now a series of positive pulses of width $\Delta\theta = \pi - \phi$ and negative pulses of width $\Delta\theta = \phi$ (for $0 \le \phi \le \pi$ as shown in the diagram) and so the phase detector characteristic is given by

$$V_p = KV_1V_2(\pi - 2\phi)/\pi \qquad (0 \le \phi \le \pi)$$

$$= KV_1V_2(\pi + 2\phi)/\pi \qquad (-\pi \le \phi \le 0) \qquad (2.5)$$

The response is now a triangular characteristic with the advantage that it is linear over the full $\pm\pi/2$ rads operating range of the detector. This allows an improvement in the capture and holding performance of the loop. From equation (2.5), the phase detector gain is now $K_p = 2KV_1V_2/\pi$ V/rad.

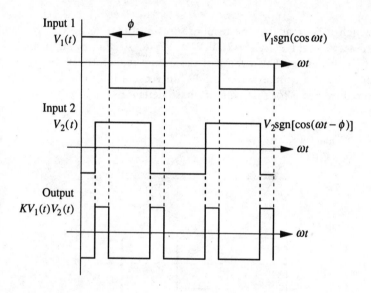

Figure 2.4 Analogue multiplier phase detector operation with square-wave inputs

From figure 2.4 it can be seen that an analogue multiplier with square-wave inputs performs an exclusive-NOR logic function. Thus, in circuits containing mostly logic components, it may be preferable to use a single exclusive-NOR (or exclusive-OR) gate as such a phase detector, as shown in figure 2.5. It should be noted that, since the output voltage of an exclusive-OR phase detector varies between the two logic levels and does not have a zero mean value, a DC offset equal to the mean of the logic '1' and logic '0' levels needs to be provided at the loop filter for the detector to function correctly. It is clear that analogue multiplier phase detectors may be based on some extremely basic circuit elements – although there are more exotic designs than the simple ones considered thus far.

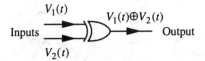

Figure 2.5 Exclusive-OR phase detector

The characteristics of the preceding analogue multiplier phase detectors used in their various modes of operation are summarised in figure 2.6. Each of these has a characteristic that repeats every 2π rads with a maximum monotonic range of π rads. If the input signals have a duty cycle which is far removed from 50% then the response will be asymmetrical, resulting in a reduction in the operating range and impaired performance in a phase-lock application. In a PLL, lock occurs when the phase detector output is zero and when the polarity of the feedback path through all the loop components is negative. Since the loop filter is usually based on an op-amp integrator with a phase inversion at DC this means that the basic loop of figure 1.1 locks on a *positive* slope of the phase detector characteristic, i.e. when $\phi_o/N - \phi_i = -\pi/2$. This quadrature phase shift from input to output is of no consequence in many applications, such as frequency discriminators, but may be important in other applications including coherent receivers. Multiplier phase detectors have the useful property that when one input signal disappears the mean output is zero. In PLLs using integrating filters (which is the majority of practical loops) this enables the loop to 'flywheel' during a temporary loss of signal and rapidly regain lock when it returns – a technique used in colour TV sets.

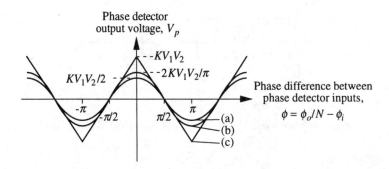

Figure 2.6 Characteristics of multiplier type phase detectors;
(a) two sinusoidal inputs, (b) one square-wave input, (c) two square-wave inputs

2.3 Harmonic locking

Analogue multiplier phase detectors with square-waves applied to one or both inputs are capable of responding to odd harmonics in addition to the fundamental frequency. This enables the loop to be locked to an odd harmonic (or sub-harmonic) of the input signal, if desired. The behaviour of an exclusive-NOR type of phase detector presented with inputs of frequency ω and 3ω is illustrated in figure 2.7. The output signal can best be thought of as a successively inverted and non-inverted version of the 3rd harmonic input signal, with the inversion being controlled by the fundamental input signal. It is clear that the phase detector output signal has a finite mean DC level that varies with the phase difference between the two applied inputs. In fact, the relative phase of the two input signals shown in figure 2.7 has been deliberately chosen to produce maximum DC response from the detector and it can be seen that this is one third of the maximum response that would be obtained from input signals of equal frequency. Thus, the phase detector gain is reduced by a factor of three.

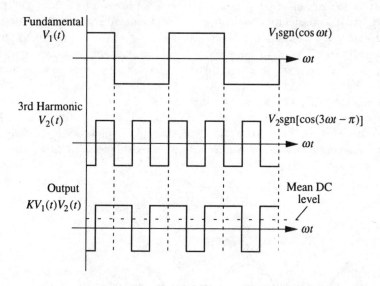

Figure 2.7 3rd harmonic operation of exclusive-NOR type phase detectors

By considering a series of diagrams such as figure 2.7 it is apparent that, for input signals of frequency ω and $n\omega$, the phase detector characteristic is triangular, as before, but the output signal level – and hence phase detector gain – is reduced by a factor of n. In practice, a further reduction in gain occurs due to the finite rise time of the square-wave edges.

Any practical phase detector has a small DC offset in its output voltage due to component tolerances and this may have an adverse effect on PLL performance, perhaps even preventing the loop from locking. The effect is much more serious in harmonically operated phase detectors since the output level is reduced by at least a factor of n and so any DC offset is much more significant. It can therefore be quite difficult to achieve high order harmonic locking in PLLs and very careful minimisation of the phase detector DC offset is required. It should also be noted that harmonic locking, far from being a useful property, may sometimes be a problem. For instance, in PLLs using frequency injection within the loop it is often possible for the loop to inadvertently lock in a harmonic mode − in which case additional measures may be required to force the loop to lock in only the desired mode.

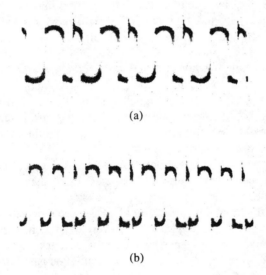

(a)

(b)

Figure 2.8 Phase detector output waveforms with (a) 3rd harmonic and (b) 5th harmonic locking

Figure 2.8 shows two waveforms obtained with a PLL circuit in which harmonic locking has been achieved. These experimental results were obtained with a phase detector similar to that of figure 2.2 in which there is a digital input which controls the inversion or non-inversion of an analogue input. In the first case, figure 2.8(a), the loop is locked to a 3rd harmonic of the analogue input signal and it is clear that there are 1.5

cycles of the analogue signal for each half-cycle of the digital signal. Similarly, in figure 2.8(b) the loop is locked to a 5th harmonic of the analogue input signal and there are now 2.5 cycles of the analogue signal for each half-cycle of the digital signal. In both cases it is evident that the mean value of phase detector output voltage is virtually zero, which is to be expected for a loop in a stable locked condition and with low DC offset voltage; this is in contrast to the example of figure 2.7 where the relative phase of the inputs is deliberately chosen to produce maximum DC output. It is also apparent that the analogue signal itself is virtually a square-wave and this is because the phase detector is preceded by a hard limiter, which is a common practice to prevent signal level changes from influencing the loop behaviour.

The phase detectors described so far produce an AC output signal in addition to the desired DC component. This alternating signal has a fundamental frequency of twice the applied signal frequency. When such phase detectors are included in PLLs, the alternating component of their outputs can tend to modulate the VCO frequency, producing unwanted sidebands on the output signal, often called *reference sidebands*. To minimise this effect, the loop natural frequency is constrained to well below the frequency at which the phase detector operates (usually around 100 times lower) in order to suppress the AC component of the phase detector output. Example 3 is designed specifically to illustrate this point. However, there is a different type of multiplying phase detector known as a sample-and-hold detector which, in principle, produces no alternating components in its output. As its name implies, it is based on a sample and hold circuit in which one phase detector input is a digital signal, the edges of which are used to determine the sampling instants of the other sinusoidal input.

Figure 2.9 shows a sample and hold phase detector in which a digital input signal is used to sample a non-inverted analogue input on rising edges and an inverted analogue input on falling edges. This is what might be called a *full-wave* sample and hold circuit compared to a *half-wave* circuit which would sample on either the rising edge *or* the falling edge. For a sinusoidal analogue input the phase detector has a sinusoidal characteristic similar to that of figure 2.6, except that the phase axis is shifted by 90° and so the phase detector output is zero when the two input signals are in phase. For non-sinusoidal analogue inputs the characteristic may be very different and possibly unusable; for instance a square-wave analogue input would produce a square-wave characteristic which could only indicate whether the two inputs are broadly in phase or in anti-phase. The circuit is therefore best used with pre-filtering and limiting or AGC in order to maintain a consistent characteristic. The resulting output signal in principle contains no AC component, though in practice there may be a small amount of second harmonic ripple due to 'drooping' in the holding capacitor. This is the major advantage of sample-and-hold detectors – they enable relatively high loop natural frequencies to be used without producing severe reference sidebands. A digital implementation of this type of phase detector is very easy to achieve by digitising the analogue input and holding its instantaneous value in a latch which is clocked on the edges of the digital phase detector input. This alleviates the problems of capacitor drooping and DC offset drifts, but introduces the possibly undesirable effects of quantisation errors.

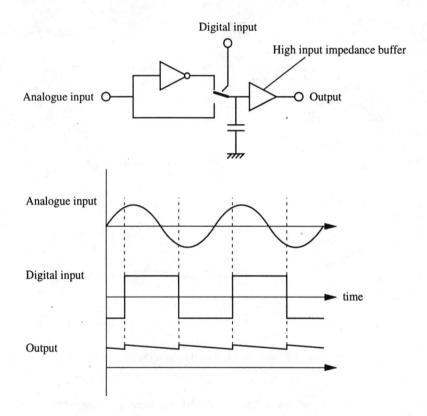

Figure 2.9 Sample-and-hold phase detector

2.4 Sequential phase detectors

We shall now look at two common types of sequential phase detector. Probably the simplest example is the RS flip-flop phase detector which is illustrated in figure 2.10. Its operation is very easy to understand by considering two digital input signals of phase difference ϕ. The flip-flop is set on the rising edge of input 1 and is reset on the rising edge of input 2. The duty cycle of the flip-flop output and the associated mean DC level is thus an indication of the phase difference between the two input signals.

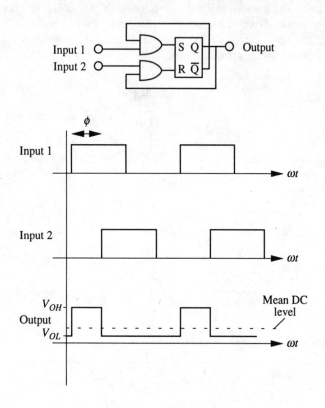

Figure 2.10 RS flip-flop phase detector and timing diagram

The mean DC output level varies linearly between the logic '0' output voltage, V_{OL}, when $\phi = 0$ to the logic '1' output voltage, V_{OH}, when $\phi = 2\pi$ rads and this characteristic is plotted in figure 2.11. In practice, a DC offset of $(V_{OL} + V_{OH})/2$ would need to be introduced at the loop filter in order to give a symmetrical output voltage swing and a PLL incorporating this phase detector would lock at a phase difference, ϕ, of π rads. This phase detector has a linear range of 2π rads – twice that of an exclusive-OR phase detector and so has an improved capture and holding characteristic. Being a sequential phase detector it responds only to the rising edges of the input signals, so their duty cycle is immaterial, although it is far less tolerant of input noise than multiplier phase detectors since a short burst of input noise may result in a prolonged and mis-timed output pulse. Also, in the absence of an input signal the mean output latches-up in one state, preventing the detector from being used in a flywheel type of PLL.

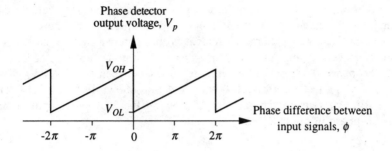

Figure 2.11 RS flip-flop phase detector characteristic

A more widely used type of sequential phase detector is the phase/ frequency detector of figure 2.12. This is available in integrated circuit form although it may be constructed from a pair of D-type flip-flops and an AND gate as shown in the diagram. Its operation is a little more subtle

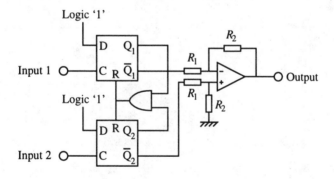

Figure 2.12 Phase/frequency detector

than the RS flip-flop phase detector. Referring to figure 2.13, assuming initially that input 1 leads input 2 by an amount ϕ, then after each rising edge of input 1 the Q_1 output is set to logic '1'. When the next rising edge of input 2 is received, for an instant both Q_1 and Q_2 outputs are set to logic '1'. This produces a pulse on the reset inputs of the flip-flops which are then both reset to logic '0'. Thus, if input 1 leads input 2 then the mean value of the Q_1 output indicates the amount of phase lead in the same way as the RS flip-flop detector, whilst the mean value of the Q_2 output is virtually V_{OL}. Conversely, if input 1 lags input 2 then the Q_2 output becomes active and indicates the amount of phase lag. By summing these two outputs in a differential amplifier, a phase detector characteristic with a linear range of 4π rads is obtained, as plotted in figure 2.14. A practical advantage of this arrangement is that compensation for offsets in the logic levels is implicit in the use of a differential amplifier. If $R_2 = R_1$ then the phase detector output voltage varies between $-(V_{OH} - V_{OL})$ and $+(V_{OH} - V_{OL})$, passing through 0 when $\phi = 0$ as shown in figure 2.14. This is a particularly useful characteristic allowing the VCO and input signals to be in-phase in a locked PLL.

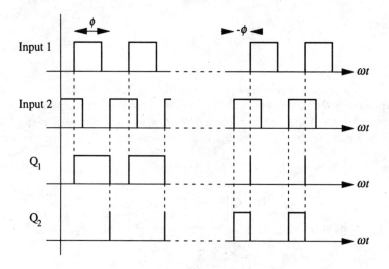

Figure 2.13 Phase/frequency detector timing diagram

It should be noted that the differential amplifier is not expected to respond to the pulses in the detector output, but merely to the DC component. Usually, a differential form of loop filter would be used to achieve the functions of both a filter and differential amplifier. The AC component of the phase detector output is a series of short pulses at the input (or reference) frequency, giving rise to reference sidebands (in a badly designed loop) at the reference frequency and its harmonics. This is in contrast to basic multiplier phase detectors where there is a much larger AC component at twice the reference frequency.

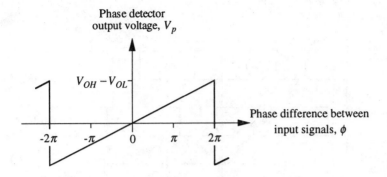

Figure 2.14 Phase/frequency detector characteristic – equal input frequencies

The most interesting feature of the phase/frequency detector is its behaviour with input signals of different frequency where, in contrast to multiplier phase detectors, it is able to discriminate between differing input frequencies. Figure 2.15 represents the behaviour of the detector with differing input frequencies. The rising edges of the two inputs and the corresponding outputs are shown. If input 1 and input 2 are of frequencies ω_1 and ω_2, respectively, and $\omega_1 > \omega_2$, then the Q_1 output is active whilst the Q_2 output is a succession of short pulses with a mean value of virtually V_{OL}. The probability that a rising edge of input 2 occurs between successive rising edges of input 1 is simply ω_2/ω_1 and its position between successive edges varies uniformly, on average being in the centre. Thus there is a $(1 - \omega_2/\omega_1)$ probability that the Q_1 output will not be reset between successive rising edges (as occurs between pulses 4 and 5 of input 1 in figure 2.15) – giving maximum mean value during that cycle – and a ω_2/ω_1 probability that the Q_1 output will be reset – giving, on average, 50% of the maximum value. The response is therefore

$$V_p = (V_{OH} - V_{OL})\left[(1 - \frac{\omega_2}{\omega_1}) + \frac{1}{2}\frac{\omega_2}{\omega_1}\right] = (V_{OH} - V_{OL})\left[1 - \frac{\omega_2}{2\omega_1}\right] \qquad (2.6)$$

Using similar arguments, for $\omega_1 < \omega_2$ the result is

$$V_p = (V_{OH} - V_{OL})\left[\frac{\omega_1}{2\omega_2} - 1\right] \qquad (2.7)$$

The characteristic, shown in figure 2.16, is interesting in that there is a discontinuity at $\omega_1/\omega_2 = 1$ indicating that, if the input frequencies differ

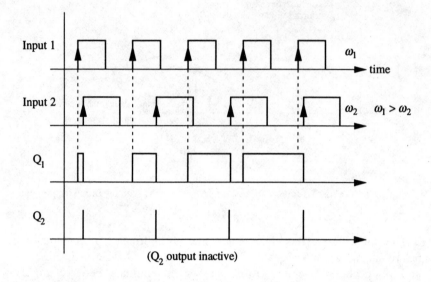

Figure 2.15 Operation of the phase/frequency detector with
non-equal input frequencies

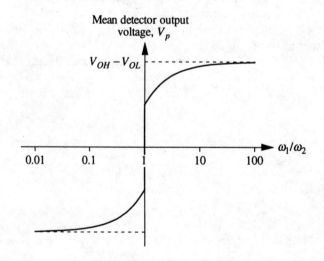

Figure 2.16 Phase/frequency detector characteristic – non-equal input frequencies

just slightly the detector produces half its maximum output voltage. This is due to the changeover between active outputs around the point where the input signals are nearly in-phase. This clear frequency discrimination characteristic makes the phase/frequency detector very useful in PLL applications where the loop is required to pull into lock from a large initial frequency offset and for this reason it is probably the most commonly used type of phase detector. The RS flip-flop phase detector also has a frequency discrimination capability, though more limited. This is the subject of example 1.

Type of phase detector Property	Multiplier	Sequential
Performance at input S/N < 10 dB	Good	Poor
Ability to 'flywheel'	Yes	No
Capability of harmonic locking	Yes	No
Capture range	Poor	Good to excellent
Tracking	Poor	Good
Optimum duty cycle of input signals	50%	Not important
AC output component frequency and level	Twice input frequency High	Input frequency Low
Phase offset in a locked loop	90°	0° or 180°

Figure 2.17 Comparison of multiplier and sequential phase detectors

Many of the relative properties of multiplier and sequential phase detectors are summarised in figure 2.17. This table is by no means exhaustive, though it does take account of the majority of the useful and interesting properties of phase detectors.

2.5 Frequency dividers

One of the most common uses of PLLs is in frequency synthesisers, where a range of output frequencies are generated from a single stable reference frequency. This simply requires the use of a variable ratio divider in the feedback path. There are other applications where only fixed dividers are required, such as in phase modulators or demodulators where a deviation beyond the range of the phase detector is needed, or in microwave frequency multiplier loops. As far as the control loop is concerned, it should be noted that frequency dividers act equally as phase dividers, so that a factor of $1/N$ must be allowed for in the loop equations.

The basic arrangement of a high frequency programmable divider is shown in figure 2.18. It uses a high frequency fixed-ratio prescaler usually made from ECL technology and capable of operation at many GHz dividing the input frequency by a fixed factor of P, followed by a programmable counter made from lower frequency TTL or CMOS technology and capable of operation at around 100 MHz dividing the frequency by a further variable factor of N, the total division ratio being NP. The duty cycle of the output signal of fixed-ratio prescalers dividing by an even number is normally 50%, whereas the duty cycle of the output signal of other prescalers and of programmable counters may be far from 50%. This fact limits the use of many programmable counters to PLLs with sequential phase detectors, which is fine in virtually all cases because of the high input S/N ratio which is guaranteed in these applications.

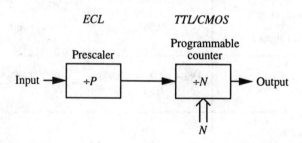

Figure 2.18 High frequency programmable divider using a fixed-ratio prescaler

The disadvantage of a PLL using this basic prescaler arrangement is that its output frequency is stepped in increments of P times the input or reference frequency. So, for a given output frequency increment, the reference frequency must be P times lower and this places more severe constraints on the loop natural frequency and hence loop performance. In particular, it increases the tuning time – which may be important in agile synthesiser applications – and it reduces the amount by which the VCO phase noise may be suppressed by locking firmly to a clean reference signal. It also reduces the spacing of reference sidebands by a factor of P which is likely to result in greater problems with close-in sidebands in communications applications.

An improved divider scheme, known as *dual-modulus prescaling*, is shown in figure 2.19. This makes use of a high frequency divider which divides by $(P + 1)$ when the modulus control input is low and P when the modulus control input is high. A special low frequency counter is used to control the division ratio of the prescaler and consists of two programmable counters and some control logic.

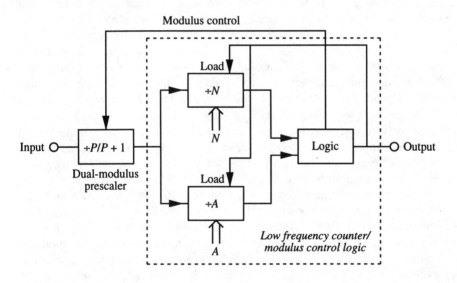

Figure 2.19 Dual-modulus prescaling

The two counters are initially loaded with the values N and A, where $N \geq A$, and the modulus control signal is low so the prescaler divides by $(P + 1)$. The counters are both decremented after each rising edge of the prescaler output until the A counter reaches zero. The modulus control signal then becomes high and the prescaler divides by P until the contents of the N counter reach zero, at which time the counters are reset and the cycle begins again. The prescaler thus divides by $(P + 1)$ for A and P for $(N - A)$ prescaler output cycles and the division ratio is therefore

$$N_t = A(P+1)+(N-A)P = NP+A, \quad \text{where } N \geq A. \qquad (2.8)$$

Thus, by varying A from 0 to $(P - 1)$, any integral value of division ratio is obtainable using this technique. This is a major advantage over the divider of figure 2.18 and, in frequency synthesiser applications, enables the output frequency to be stepped in increments of the reference frequency. Common prescaler ratios are 8/9 – allowing 3 A bits to be grouped with the N bits and treated as a single binary input and 10/11 – allowing a binary-coded-decimal representation of the division ratio.

As an example of a common arrangement, consider the case of an 8/9 prescaler. The A bits are varied between 0 and 7 and the N bits from 7 upwards (since $N \geq A$). Thus the division ratio is adjustable in unity increments from 56 upwards. This lower limit on the available division ratio is most unlikely to be a limitation in a practical synthesiser circuit where the division ratios required are likely to be very high.

The more detailed timing considerations can be understood with reference to the example of figure 2.20 where $N = 5$, $A = 3$ and $P = 4$. The N and A counters are initially loaded with values of 5 and 3, respectively, and are decremented after each rising edge of the prescaler output. When $A = 3$, 2 and 1 the prescaler divides by 5 and when $N = 2$ and 1 it divides by 4, giving a total division ratio, as expected, of 23. The output signal is derived from the short pulse used to reset the counters. It is important to note that the prescaler division ratio is determined by the state of the modulus control on the rising input edge when the prescaler output is about to become high. Thus the modulus control signal should be triggered by the prescaler output edge *prior* to the counter state in which the division ratio needs to be changed, i.e. $A = 1$ and $N = 1$. This arrangement allows the relatively low frequency counter and modulus control logic the maximum possible time in which to change the state of the modulus control – P or $(P + 1)$ periods of the input signal. In practice, look-ahead decoding is often used to further increase the speed of operation.

The division ratios of dual-modulus prescalers can be increased by using an arrangement such as that of figure 2.21. Dual modulus prescalers are usually designed with this in mind and include several modulus control inputs so that the external OR gate shown in the figure is not actually required. The prescaler is followed by a $\div 4$ switch-tail divider, the two outputs of which are used as modulus control inputs. In three of the four possible flip-flop states the modulus control is thus held high and the prescaler is forced to divide by P. However, in the state before the Q_2 output becomes high, the modulus control level is determined by the external control input and thus the prescaler division ratio of the next cycle may be P or $(P + 1)$. The circuit therefore functions as a $\div 4P/(4P + 1)$ prescaler. It is important, from the timing considerations mentioned previously, that the modulus control input is active at the time when the Q_2 output is about to become high. This point is often overlooked in data books and erroneous circuits are shown.

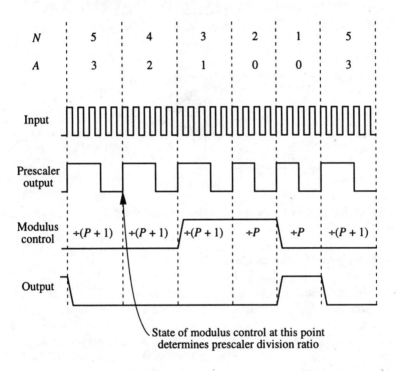

Figure 2.20 Dual-modulus prescaler timing diagram; $N = 5$, $A = 3$, $P = 4$

A further type of frequency divider which has applications in agile synthesisers, is known as a *fractional-N* divider. This, as the name implies, is capable of dividing by a fractional number. In reality however, it actually toggles between two successive integers to provide a mean division ratio somewhere in between. This type of divider, discussed in greater detail in chapter 8, requires quite elaborate circuitry because it produces a large amount of sawtooth modulation at sub-reference frequencies which is reduced by applying a similar compensating waveform elsewhere in the loop.

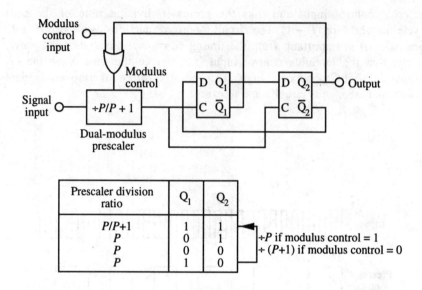

Figure 2.21 Extension of the dual-modulus prescaler division ratio to $4P/(4P + 1)$

2.6   Voltage-controlled oscillators (VCOs)

VCOs are electronically tunable oscillators in which the output frequency, ω_o, is dependent on the value of an applied tuning voltage, V_t. They are realised in many forms from RC multivibrators at low frequencies to varactor and YIG-tuned oscillators at higher frequencies. As far as the loop filter design is concerned, the most important property of VCOs is their tuning characteristic. The slope of this characteristic, the *VCO gain*, is a further factor to be included in the loop equations, in addition to the phase detector gain, K_p, and divider ratio, $1/N$, and is defined as

$$K_v = \frac{d\omega_o}{dV_t} \quad \text{rad/s/V}$$

$$(2.9)$$

where ω_o is the output frequency and V_t is the tuning voltage. A typical tuning characteristic is shown in figure 2.22, where it should be appreciated that units of rad/s/V would normally be used in loop calculations for compatibility with the units of the phase detector gain. In a typical varactor-tuned LC oscillator, as the tuning voltage is increased, the VCO frequency initially increases rapidly and then increases more gradually. The VCO gain therefore decreases as the operating frequency is increased. Typically, the VCO gain may be expected to vary by a factor of two in a varactor-tuned oscillator having a tuning range of around half an octave. This is compounded in frequency synthesiser applications by the $1/N$ variation in frequency divider gain so that the overall loop gain may vary quite considerably. In practice, some means of compensation would usually be provided for significant variations in loop parameters with operating frequency. This may take the form of a non-linear DC amplifier placed before the VCO, or a loop filter with gain related to the division ratio.

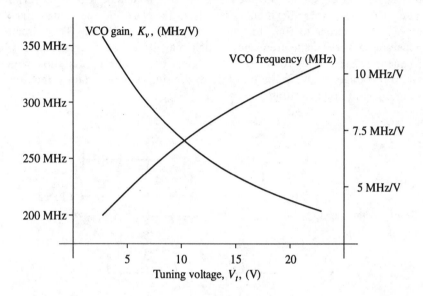

Figure 2.22 A typical VCO characteristic

A further important property of VCOs in PLL applications is their degree of phase noise purity (although amplitude noise is not as important since this may be removed by limiting). This is a complex subject, although it is fair to say that spectral purity is dependent largely on the Q of the resonant element in the oscillator. In a loop having a wide bandwidth, a significant improvement above the residual VCO phase noise may be obtained by locking to a stable reference frequency – derived, perhaps, from a quartz crystal oscillator.

Some typical VCO circuits are given in figure 2.23. Figure 2.23(a) shows a VHF VCO which is based on a Colpitts oscillator in which the tapped capacitors are replaced by a pair of varactor (otherwise known as varicap) diodes. By increasing the reverse bias on these diodes, their capacitance falls and so the frequency of oscillation increases. The circuit shown has a frequency adjustment range of around half an octave. It should be noted that the variation of frequency with control voltage is somewhat non-linear in a wide-range VCO such as this. There are two compounded reasons for the non-linearity: firstly the varactor capacitance varies non-linearly with bias voltage and secondly the frequency of oscillation varies non-linearly with the varactor capacitance.

Figure 2.23(b) shows a voltage controlled crystal oscillator (VCXO). This is based on a standard crystal oscillator circuit comprising a parallel resistance and crystal in the feedback path of a Schmitt trigger invertor gate. However, varactor diodes are used in place of the usual fixed capacitors in order to fine tune the frequency of oscillation. This circuit generates a very stable frequency which may be adjusted over a narrow frequency range of approximately ±25 ppm. The frequency variation with control voltage of this circuit is highly non-linear and certainly requires some form of compensation.

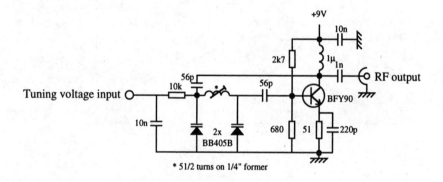

Figure 2.23(a) A varactor-tuned Colpitts VCO

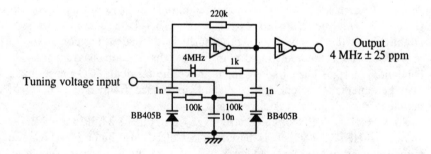

Figure 2.23(b) A voltage-controlled crystal oscillator

There are a vast number of VCO circuits, some based on op-amp or comparator techniques offering high linearity and wide frequency range. Unfortunately these tend to operate at low frequencies and have relatively poor phase noise performance.

2.7 Loop filters

Operation of the PLL of figure 1.1 may be represented by the signal flow graph shown in figure 2.24 in which each block in the loop is replaced by an arrowed line between two nodes and each node is labelled with the value of the parameter at that point, for example frequency at the output of the VCO. Here the filter transfer function is represented, using Laplace notation, by $G(s)$ and a $1/s$ term is included to translate the VCO output frequency, ω_o, into phase, ϕ_o. For analytical convenience, negative feedback is included by means of the negative polarity given to the frequency divider although, in practice, this negative term will actually arise from the loop filter or phase detector characteristic. The closed-loop transfer function may be found by applying Mason's rule which, for a single loop control system such as this, reduces to the following simple formula:

$$Closed\text{-}loop\ transfer\ function\ =\ \frac{output}{input}\ =\ \frac{path\ gain\ from\ input\ to\ output}{1\ -\ loop\ gain}$$

(2.10)

Applying this to the signal flow graph of figure 2.24 gives the result,

$$\frac{\phi_o(s)}{\phi_i(s)} = \frac{\dfrac{K_pK_vG(s)}{s}}{1+\dfrac{K_pK_vG(s)}{Ns}}$$

(2.11)

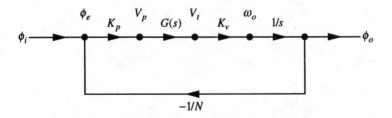

Figure 2.24 Signal flow graph representation of a PLL

Loops are categorised according to their *order* and *type*, the definitions being derived from classical control theory. The order of a loop is defined by the highest power of s in the denominator of the closed-loop transfer function shown above and the type of loop is defined by the number of perfect integrators within the loop. All loops are at least type I because of the integrating action of the VCO.

The simplest loop is obtained when $G(s) = 1$ for which the closed-loop transfer function is

$$\frac{\phi_o(s)}{\phi_i(s)} = \frac{N}{\dfrac{N}{K_p K_v}s + 1} \tag{2.12}$$

This is clearly a first order type I loop and the loop transfer function is simply a first order low-pass response, identical to that of the simple filter shown in figure 2.25, with a time constant of $N/(K_p K_v)$. This loop is of very limited practical use since there is no filtering of the phase detector output and so reference sidebands are likely to be very high. An improvement can be obtained by using a passive low-pass filter, but this also has limitations such as there being a finite phase offset necessary to support a steady-state VCO control voltage and possible acquisition and tracking problems. More useful loop filters fulfil a combination of low-pass and integrator properties, the latter function requiring the use of an active device such as an op-amp. In practice, most designs include active loop filters to take advantage of the improved dynamic performance and reference sideband suppression. Passive filters are, however, useful in very fast applications where it is difficult to implement a stable active loop filter.

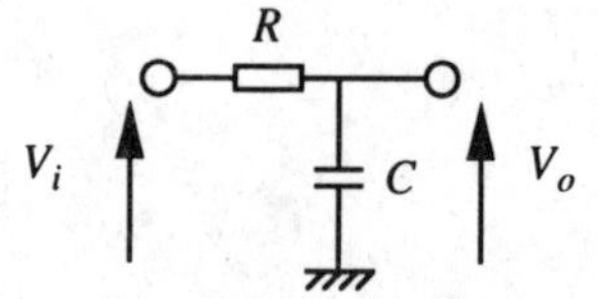

$$\frac{V_o(s)}{V_i(s)} = \frac{1/sC}{R + 1/sC} = \frac{1}{sCR + 1} = \frac{1}{\tau s + 1}$$

Figure 2.25 Simple first order low-pass filter

3 Loop Basics

3.1 Design principles

PLLs, being a form of control system, may be designed using standard control techniques in which the loop filter design, in conjunction with the other loop components, determines the characteristics of the system. Figure 3.1 shows a selection of loop filter designs ranging from a passive RC filter, figure 3.1(a), to an active 3rd order filter, figure 3.1(d). Generally, it may be said that higher order active filters provide a better overall compromise in loop performance, although there are occasions when a simple passive filter may be more appropriate. An example of such an occasion is in the design of very fast loops where active elements with non-ideal behaviour or parasitic delays, such as track lengths, may introduce unwanted phase shifts around the loop, resulting in instability. Such high speed loops may be used, for example, in microwave modulation of fibre-optic communication links. In such cases, a passive loop filter may be the only option – since it will have a more controllable phase characteristic than even a simple active loop filter. A more recent development is the use of all-digital loops where the phase detector, loop filter and VCO may be made entirely from digital components; this subject is covered in more detail in chapter 8.

One of the most useful and popular designs of loop filter, producing a second order type II response, is shown in figures 3.1(b) and 3.1(c). The single-ended version of figure 3.1(b) is suitable for use with multiplier phase detectors or sequential phase detectors with single-ended (such as tri-state) outputs, whereas the differential form of figure 3.1(c) is suitable for use with phase/frequency detectors having two complementary outputs. This design of loop filter functions as an integrator at low frequencies with a DC gain equal to that of the op-amp. Since this gain is usually very high (around $10^5 - 10^6$) it may be regarded as a perfect integrator at low frequencies, for which the filter transfer function is

$$G(s) = \frac{R_2 + 1/sC}{R_1} \tag{3.1}$$

Substituting this into equation (2.11) yields the closed-loop transfer function,

$$\frac{\phi_o(s)}{\phi_i(s)} = \left[\frac{K_p K_v R_2}{R_1}\right] \frac{s + 1/CR_2}{s^2 + \dfrac{K_p K_v R_2}{NR_1}s + \dfrac{K_p K_v}{NR_1 C}}$$

This result, in fact, describes a damped simple harmonic motion (SHM) behaviour which applies to many electrical and mechanical systems including, for example, a mass on the end of a spring and the suspension components of a motor vehicle.

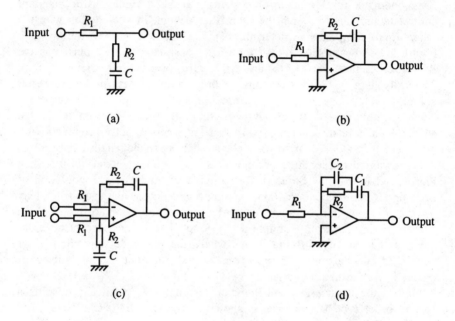

Figure 3.1 A selection of loop filters. (a) Second order type I; (b) Second order type II; (c) Second order type II with differential inputs; (d) Third order type II

The standard response of such a system is as follows:

$$\frac{\phi_o(s)}{N\phi_i(s)} = \frac{2\zeta\omega_n s + \omega_n^2}{s^2 + 2\zeta\omega_n s + \omega_n^2} \tag{3.2}$$

where ω_n is the natural frequency and ζ the damping factor. By comparison with the closed loop transfer function obtained for a PLL with a second order type II loop it is evident that the loop natural frequency, ω_n, and damping factor, ζ, are given by

$$\omega_n = \sqrt{\frac{K_p K_v}{NR_1 C}} \text{ and } \zeta = \frac{R_2}{2}\sqrt{\frac{K_p K_v C}{NR_1}} = \frac{\omega_n R_2 C}{2} \tag{3.3}$$

These are the design equations for this particular loop filter, the loop properties depending entirely on the choice of the two parameters ω_n and ζ. In general, ω_n determines the cut-off frequency of the response and ζ determines the shape of the characteristic, $\zeta = 1$ being the case for critical damping. A value for ζ of between 0.5 and 1 is normally used, with 0.707 being a popular design choice because it gives rise to a Butterworth polynomial in the denominator of equation (3.2). Because of the widespread use of the second order type II loop and its relative simplicity of analysis, this loop will be considered extensively (though by no means exclusively) throughout this book.

The passive loop filter design of figure 3.1(a) clearly does not have an integrating function and therefore gives rise to a second order type I loop. This is sometimes referred to as a low gain loop because of the absence of any high-gain active element. Its closed loop transfer function can be found in an identical manner to before, resulting in

$$\frac{\phi_o(s)}{N\phi_i(s)} = \frac{(2\zeta\omega_n - N\omega_n^2/K_p K_v)s + \omega_n^2}{s^2 + 2\zeta\omega_n s + \omega_n^2} \tag{3.4}$$

with

$$\omega_n = \sqrt{\frac{K_p K_v}{N(R_1 + R_2)C}} \text{ and } \zeta = \frac{\left[R_2 C + N/K_p K_v\right]}{2}\sqrt{\frac{K_p K_v}{N(R_1 + R_2)C}} \tag{3.5}$$

The third order loop design of figure 3.1(d) is considered in detail in chapter 7.

3.2 Stability

Before beginning an analysis of loop stability it may be worth reviewing the stability criteria summarised in appendix A. To determine the stability of the loop, the loop gain $K_p K_v G(s)/Ns$ must be considered which, for a second order type II loop, is

$$loop\ gain = \frac{K_p K_v G(s)}{Ns} = \frac{\phi_o/N\phi_i}{1 - \phi_o/N\phi_i} = \frac{2\zeta\omega_n s + \omega_n^2}{s^2} \tag{3.6}$$

The polar plot of this function, shown in figure 3.2, indicates that the gain tends to $-\infty$ at DC, tends asymptotically to $-0j$ at high frequencies and at all times lies in the third quadrant (i.e. with negative real and imaginary components). However, the loop filter is based on an imperfect integrator which eventually functions as a high gain amplifier at very low frequencies, thus bringing the eventual low frequency response to $-\infty j$. Since the phase never quite reaches $-180°$ then the gain margin is infinite and so analysis is restricted to the phase margin only. The modulus of the loop gain is

$$|loop\ gain| = \frac{\sqrt{1 + 4\zeta^2(\omega/\omega_n)^2}}{(\omega/\omega_n)^2}$$

and this falls from infinity at DC to zero at frequencies well above ω_n, passing through unity at a frequency ω' given by

$$(\omega'/\omega_n)^4 = 1 + 4\zeta^2(\omega'/\omega_n)^2$$

which results in a quadratic equation in $(\omega'/\omega_n)^2$ with the solution

$$(\omega'/\omega_n)^2 = 2\zeta^2 + \sqrt{4\zeta^4 + 1}$$

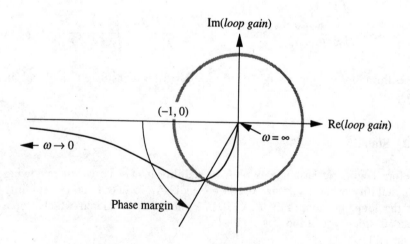

Figure 3.2 Polar plot of loop gain for the second order type II PLL

The argument of the loop gain is

$$\arg(\textit{loop gain}) = \tan^{-1}(2\zeta\,\omega/\omega_n) - 180°$$

and so the phase margin, being defined as the difference between the argument of the loop gain and $-180°$ at the frequency where the loop gain is unity, is

$$\theta = \tan^{-1}(2\zeta\,\omega/\omega_n) = \tan^{-1}\left[2\zeta\sqrt{2\zeta^2 + \sqrt{4\zeta^4 + 1}}\right] \qquad (3.7)$$

The phase margin for some commonly used damping factors is shown in table 3.1. The loop is stable, in principle, for all damping factor values – although at very low damping factors stability is marginal, with a characteristic oscillatory ringing response to transients, whilst at high damping factors the loop response is very sluggish. For the usual range of damping factors, between 0.5 and 1, the phase margin lies healthily between 51.8° and 76.3°.

Table 3.1 Phase margin versus damping factor for a second order type II PLL

Damping factor ζ	Phase margin (°)
0	0
0.5	51.8
0.707	65.5
1	76.3
∞	90

A practical circuit for a frequency synthesiser, shown in figure 3.3, demonstrates many of the concepts featured so far in chapters 2 and 3. The circuit generates output frequencies around 1.5 GHz using a monolithic VCO (the VTO9120) which is buffered to isolate the VCO from changes in the output load which could otherwise pull the VCO frequency and degrade the overall phase noise purity. The VCO output is initially divided by 4 by a fixed ratio ECL prescaler (SP8712B) and is then applied to a ÷32/33 prescaler formed from a ÷8/9 prescaler (MC12011) in conjunction with a switched-tail modulo 4 counter using two D-type flip-flops (MC10131). The final output of this divider chain is at a frequency of

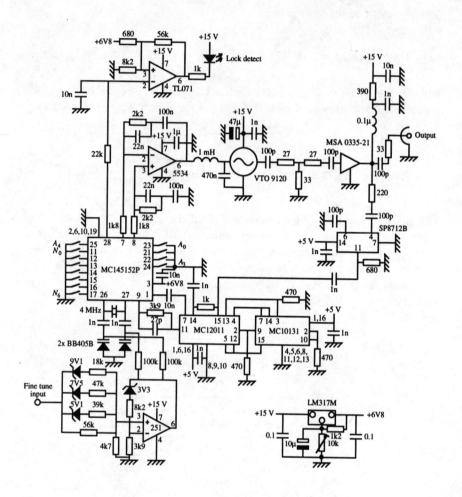

Figure 3.3   Circuit diagram of an L-band PLL frequency synthesiser

around 12 MHz which is then presented to a PLL chip (the MC145152P) made from CMOS technology. The MC145152P is capable of reliable operation at up to 25 MHz and in addition to the dual-modulus control circuitry contains a reference divider, a phase/frequency detector and a lock detector. The reference divider divides the 4 MHz crystal frequency by a factor of 256 to a final reference frequency of just 15.625 kHz. However, because of the fixed ÷4 prescaler, the smallest output frequency increment is 62.5 kHz. The outputs of the phase/frequency detector are applied to the inputs of a differential third order type II loop filter (see chapter 7) which provides the VCO tuning voltage. The loop natural frequency in this design is 1.5 kHz which is a significant fraction of the reference frequency and would normally result in high reference sidebands. To counter this effect, additional LC filtering is included between the loop filter and VCO though, as shown in chapter 7, it is important to ensure that this additional filtering does not significantly degrade loop stability. The circuit, unusually, has an electronically tunable reference frequency by means of reverse-biased varactor diodes around a 4 MHz crystal oscillator. Because of the inherent non-linearity of this fine-tuning, a 4-stage DC amplifier is used as compensation − resulting in a reasonably linear output tuning range of ±55 kHz.

3.3 Transient response

Having found the closed-loop transfer function, $\phi_o(s)/\phi_i(s)$, it is possible to obtain either the frequency response to sinusoidal input components or the transient response to a given input change. The former matter is of great interest in the use of PLLs for modulation and demodulation purposes and also when considering the noise performance of the loop and is considered in some detail in chapters 4 and 5. In this section, standard Laplace transform techniques are employed to arrive at some useful and interesting results concerning phase and frequency transients in PLLs.

3.3.1 Loop response to a step change in phase

Perhaps the most obvious case of transient effects is that of a step change in the input phase of a previously locked PLL. This situation could arise due to a phase change in the reference source or due to a sudden change in the VCO phase as a result of power supply transients. The situation may readily be analysed by considering the Laplace transform of an input phase step of magnitude $\Delta\phi$ occurring at $t = 0$,

$$\phi_i(t) = \Delta\phi; \quad \phi_i(s) = \frac{\Delta\phi}{s}$$

to which the loop response is

$$\frac{\phi_o(s)}{N} = \Delta\phi\left[\frac{2\zeta\omega_n s + \omega_n^2}{s(s^2 + 2\zeta\omega_n s + \omega_n^2)}\right]$$

Some standard Laplace transforms are given in appendix G. By resolving into partial fractions and taking the inverse Laplace transform, the corresponding phase transient may be found; however, the form of the result depends on the range of values of ζ. We shall consider the case of $\zeta = 1$ since it represents the case of critical damping – a realistic, in fact, ideal choice. Thus the loop response for unity damping factor is

$$\frac{\phi_o(s)}{N} = \Delta\phi\left[\frac{2\omega_n s + \omega_n^2}{s(s + \omega_n)^2}\right]$$

and, resolving into partial fractions, this is equivalent to

$$\frac{\phi_o(s)}{N} = \Delta\phi\left[\frac{1}{s} - \frac{1}{s + \omega_n} + \frac{\omega_n}{(s + \omega_n)^2}\right] \tag{3.8}$$

Taking the inverse Laplace transform, the loop response in the time domain is therefore

$$\frac{\phi_o(t)}{N} = \Delta\phi\left[1 - (1 - \omega_n t)e^{-\omega_n t}\right] \tag{3.9}$$

and so the phase *error* transient experienced at the phase detector is given by

$$\phi_e(t) = \Delta\phi - \frac{\phi_o(t)}{N} = \Delta\phi(1 - \omega_n t)e^{-\omega_n t} \tag{3.10}$$

and this is plotted in figure 3.4. Naturally enough, the phase error is initially $\Delta\phi$ at the instant when the phase step is applied and then rapidly decays, becoming negative when $\omega_n t \geq 1$ and afterwards gradually tending to zero. This describes the expected critically-damped response and so $\zeta = 1$ represents an optimum choice of damping factor for this loop design. It should be noted that the output phase error (i.e. the VCO phase error) is N times that present at the phase detector, in other words,

$$\Delta\phi_o(t) = N\phi_e(t) \tag{3.11}$$

For a given phase step it is possible to accurately predict the corresponding loop response provided that the phase step is within the linear range of the phase detector. In a frequency synthesiser, where the division ratio may be very high, the allowable phase step of the VCO may be quite significant and yet still be within the linear range of the loop.

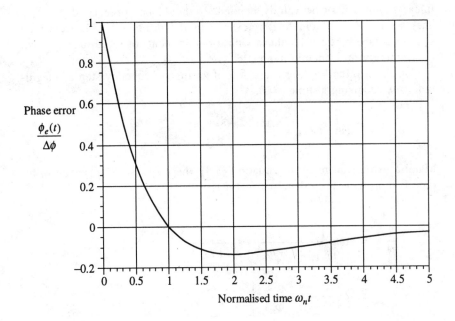

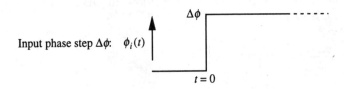

Figure 3.4 Response of a second order type II PLL ($\zeta = 1$) to a step change in input phase.

3.3.2 Loop response to a step change in frequency

This case is of particular relevance to frequency synthesisers, where a change in the division ratio causes a phase error transient at the phase detector which results in the adjustment of VCO frequency to a new steady-state value and an eventual nulling of the phase error. It is possible to precisely analyse both phase and frequency transients using linear methods, provided that the frequency step is small enough to avoid taking the phase detector beyond its linear operating range. The following analysis determines the form of the phase and frequency transients at the loop output and also the maximum allowable frequency step in order that a linear analysis may be validly applied. It should be appreciated that this analysis is of great practical relevance in the design of PLL frequency synthesisers – particularly those with a requirement for rapid tuning – and is not merely of academic interest!

Considering the Laplace transform of an input frequency step of magnitude $\Delta\omega$, occurring at time $t = 0$,

$$\phi_i(t) = \Delta\omega t; \quad \phi_i(s) = \frac{\Delta\omega}{s^2}$$

Making use of the result of equation (3.8), the response of a second order type II loop with unity damping factor is

$$\frac{\phi_o(s)}{N} = \Delta\omega\left[\frac{1}{s^2} - \frac{1}{s(s+\omega_n)} + \frac{\omega_n}{s(s+\omega_n)^2}\right]$$

$$\equiv \Delta\omega\left[\frac{1}{s^2} - \frac{1}{(s+\omega_n)^2}\right]$$

and taking the inverse Laplace transform, the loop response in the time domain is thus

$$\frac{\phi_o(t)}{N} = \Delta\omega t\left[1 - e^{-\omega_n t}\right] \tag{3.12}$$

The phase error transient experienced at the phase detector is equal to the difference between the input phase, $\Delta\omega t$, and the fed-back output phase,

$$\phi_e(t) = \Delta\omega t - \frac{\phi_o(t)}{N} = \Delta\omega t e^{-\omega_n t} \equiv \frac{\Delta\omega}{\omega_n}\omega_n t e^{-\omega_n t} \tag{3.13}$$

and this is plotted in figure 3.5(a). The phase error is initially zero (as expected since the input phase shift is proportional to time), rises to a peak of around $0.37(\Delta\omega/\omega_n)$ at $\omega_n t = 1$ and then gradually tends to zero as the loop settles to its new operating frequency. It is easy to establish the peak phase error by finding the turning point of equation (3.13):

$$\frac{d\phi_e(t)}{d(\omega_n t)} = \frac{\Delta\omega}{\omega_n}\left[e^{-\omega_n t} - \omega_n t e^{-\omega_n t}\right] = 0 \quad \text{when } \omega_n t = 1$$

and so the peak phase error is

$$\phi_e(\text{peak}) = \frac{\Delta\omega}{e\omega_n} \tag{3.14}$$

This is an important result because it determines the maximum frequency step which is permissible in a frequency synthesiser with the loop still operating in a linear manner. For instance, assuming that a phase/frequency detector is used (with an operating range of $\pm 2\pi$ rads) then the maximum allowable input frequency step for the loop to operate in a linear fashion is

$$\Delta f_i(\text{max}) = 2\pi e f_n$$

and this scales by a factor of N (the divider ratio for $t \geq 0$) so that the maximum allowable output frequency step is

$$\Delta f_o(\text{max}) = 2\pi e N f_n \tag{3.15}$$

which may be quite a significant change in frequency. The phase error expression of equation (3.13) may be differentiated to find the frequency error transient at the phase detector arising from a frequency step,

$$\omega_e(t) = \frac{d\phi_e(t)}{dt} = \Delta\omega(1 - \omega_n t)e^{-\omega_n t} \tag{3.16}$$

which is of a similar form to equation (3.10) and is plotted in figure 3.5(b). The frequency error is initially $\Delta\omega$ at the instant when the frequency step is applied and then rapidly decays, becoming negative when $\omega_n t \geq 1$ and afterwards gradually tending to zero. In many practical cases the frequency step is small enough to allow linear loop operation, in which case circuit behaviour, including loop settling time, may be very accurately predicted from the analytical result given here and plotted in figure 3.5(b).

As an example of the use of the ideas described in this section, consider the case of a PLL frequency synthesiser generating frequencies between 150 MHz and 175 MHz by locking to a 25 kHz reference using a programmable divider with division ratios adjustable between 6000 and 7000. A realistic design would be a second order type II loop with unity damping factor and a natural frequency of 250 Hz, employing a phase/frequency detector. The worst case frequency step occurs when the output frequency is changed from 175 MHz to 150 MHz by changing the divider ratio from

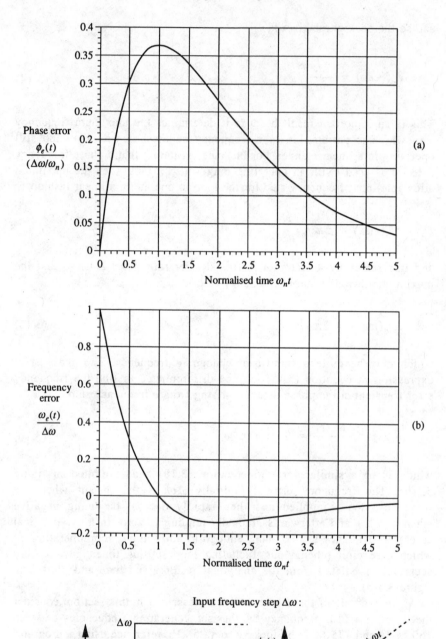

Figure 3.5 Response of a second order type II PLL ($\zeta = 1$) to a frequency step

7000 to 6000. The frequency step observed at the phase detector in this situation is 25 MHz/6000 = 4.167 kHz and, from equation (3.14), the peak in the phase error transient is $4167/250e$ Hz = 6.14 radians or 351°. This is just within the operating range of the phase/frequency detector and so loop operation during the change in channel frequency will be described by the preceding linear analysis with an output frequency error (from equation (3.16)) given by

$$\Delta f_o(t) = N f_e(t) = 25(1-1571t)e^{-1571t} \text{ MHz}$$

This equation may be used to find when the synthesiser has settled to within a certain value of its final frequency; for example, the time taken to reach within 1 kHz of the final frequency is approximately 8 ms. This example demonstrates that a simple and exact linear analysis is applicable over a remarkably wide frequency step. Beyond this linear range, operation is a little slower and only predictable using numerical methods – often involving a number of approximations. Example 9 applies to the performance of a third order type II loop under exactly the same conditions and makes for an interesting comparison between the transient responses of the two filter designs.

3.4 Measurement techniques

It is relatively easy to measure the performance of the four basic loop components in isolation. In the case of the frequency divider, the task is almost trivial. The division ratio can be ascertained by comparing the frequency of an input signal with the associated output signal – which simply requires a frequency counter with sufficient resolution.

The loop filter response is also very easy to establish by a simple measurement using an audio spectrum analyser. If this item of test equipment is not available then a manually swept audio oscillator, voltmeter and oscilloscope may be used to measure the amplitude and approximate phase characteristics of the filter.

The main item of interest regarding VCO performance is the output frequency versus control voltage characteristic. This can easily be obtained with the help of a frequency counter, voltmeter and variable power supply and then plotted, the slope of the tuning characteristic giving the VCO gain which, of course, is an important parameter in the loop filter design (see figure 2.22). VCO spectral purity may also be of concern and this may be established in a number of ways. The most accurate method is to use a modulation meter to separately measure the residual phase noise and amplitude noise profiles. A different and more approximate method may be employed if the VCO output is amplitude limited or if it is suspected that phase noise is dominant. The spectral profile of the VCO may be measured on a spectrum analyser and, assuming that the total VCO phase noise is relatively small (< 15° approximately), a narrow band phase modulation approximation may be made in which – as it is shown in appendix B – the phase noise spectral density (in dBrad/Hz) is of the same shape as the power spectral density (in dBc/Hz) but a factor of 6 dB higher.

The characteristic of digital phase detectors is known quite accurately from measurement of the output logic levels. For instance, a dual D-type phase/frequency detector made from standard CMOS logic operating from a 15 V supply would have a phase detector gain of $15/2\pi$ V/rad. If the same device were made from any general logic family, the phase detector gain would be $(V_{OH} - V_{OL})/2\pi$ V/rad. Analogue phase detectors, however, require a different measurement strategy because their behaviour varies with input signal levels (and shapes) and is more sensitive to component tolerances. One approach is to build the complete PLL and to open-loop the system by disconnecting the loop filter output and applying an externally adjustable voltage to the VCO input in order to sweep the VCO frequency. By adjusting the VCO frequency so that it is close to the reference frequency and viewing the resulting low frequency beat note at the phase detector output, the phase detector characteristic may be obtained from which the phase detector gain may be inferred.

It is often of greater value to measure the loop performance as a whole, since the effects of any errors in the estimation of loop component parameters may be cumulative and there may be other uncertainties in the circuit design which have been overlooked. Also, in situations where the loop is presented as a 'black box', closed loop measurement may be the only option. Figure 3.6 shows an experimental arrangement designed to measure the closed loop transfer function, $\phi_o(s)/\phi_i(s)$, in both amplitude and phase. It requires an audio network analyser which provides a swept frequency sinusoidal signal to the phase modulation input of an RF signal generator. The resultant phase modulated carrier is applied to the PLL under test, the output of which is demodulated by a modulation meter and then compared by the network analyser with the original modulation. The PLL transfer function in both amplitude and phase can therefore be recorded over a range of modulating frequencies. In order to eliminate the imperfect behaviour of the signal generator and modulation meter, a calibration loop should be used to produce a response which may be subtracted (in both amplitude and phase) from the measured results with a PLL to produce calibrated data.

Measurement of the closed loop transfer function may be a goal in itself – since it is a direct indicator of many aspects of loop performance. However, it may be of interest to derive the loop gain from this measurement to establish, for instance, the loop filter design – order, type, natural frequency, damping factor and so on. This is relatively easy to do once the

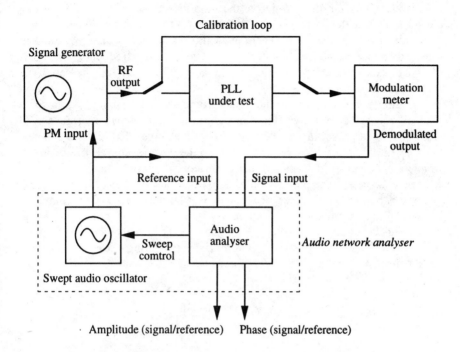

Figure 3.6 Experimental arrangement to measure the closed loop transfer function

complex closed loop characteristic is obtained. Recalling from the signal flow graph of figure 2.24 that the closed loop transfer function may be expressed as

$$\frac{\phi_o(s)}{N\phi_i(s)} = \frac{\dfrac{K_p K_v G(s)}{Ns}}{1 + \dfrac{K_p K_v G(s)}{Ns}}$$

then, by simple rearrangement of this equation, the loop gain may be inferred from the complex closed loop response experimentally found from an arrangement along the lines of figure 3.6,

$$\frac{K_p K_v G(s)}{Ns} = \frac{\dfrac{\phi_o(s)}{N\phi_i(s)}}{1 - \dfrac{\phi_o(s)}{N\phi_i(s)}} \qquad (3.17)$$

It is clear that, in order to derive the loop gain from measurement of the closed-loop transfer function, the divider ratio, N, must be known. If we are dealing with a black box where even this parameter is unknown then it is possible to accurately estimate N by noting the low frequency tendency of the closed-loop characteristic which, as will be seen in chapter 4, is indeed the factor N!

Dynamic measurement of the tuning time of a frequency synthesiser may be accurately made by the simple arrangement shown in figure 3.7. Here, the synthesiser under test is made to change channel frequencies from f_1 to f_2, say, whilst its output is mixed with a source of identical frequency, f_2, obtained from a signal generator which is locked to the same reference as the frequency synthesiser. Thus the synthesiser and signal generator output frequencies are coherent (though their phases may be different) and so the mixer output consists of a beat note accurately indicating the proximity to phase lock. The mixer output may be observed on a digital oscilloscope, with appropriate triggering, from which the tuning time may be accurately estimated.

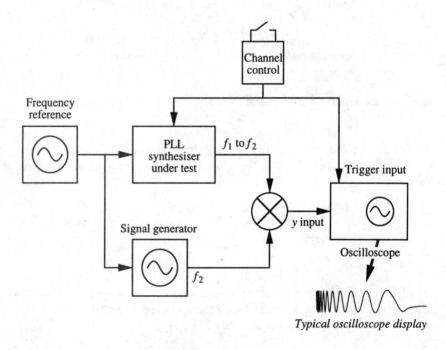

Figure 3.7 Experimental method for measuring synthesiser tuning time

3.5 Phase-plane representation

We have already seen, in section 3.3, an example of the calculation of loop phase error and frequency error transients arising from a step change in input phase or frequency. By using standard Laplace analyses, the form of a number of transients was derived and plotted (in figures 3.4 and 3.5) for a second order type II loop with unity damping factor. Taking figure 3.5, for example, both the phase and frequency transients resulting from a step change in loop input frequency are plotted as a function of time and provide a clear representation of the loop response to the original input excitation. However, there is another way of displaying the results of such an analysis which is quite widely used and is known as the *phase-plane* representation. To produce such a phase-plane plot, the frequency error, $\dot{\theta}_e(t)$, is plotted against the phase error, $\theta_e(t)$, providing a locus which describes the frequency/phase loop error variation for values of time, t, from 0 to ∞. Phase-plane plots are useful for representing transient behaviour – particularly under non-linear conditions – and for a number of other dynamic and/or non-linear situations, such as chaotic behaviour which can occur under certain conditions in PLLs.

An example of the phase-plane representation of a frequency step in a linear PLL may be arrived at simply by treating the phase and frequency transient equations plotted in figure 3.5 as parametric equations of the variable $\omega_n t$:

$$\phi_e(\omega_n t) = \frac{\Delta\omega}{\omega_n}\omega_n t e^{-\omega_n t} \tag{3.18a}$$

and

$$\omega_e(\omega_n t) = \Delta\omega(1 - \omega_n t)e^{-\omega_n t} \tag{3.18b}$$

These equations are plotted in figure 3.8 and provide a phase-plane plot corresponding to the two time-domain plots of figure 3.5. Referring to this plot, at time $t = 0_-$, immediately before the frequency step is applied, both frequency and phase errors are zero, indicating the case of a settled, locked loop. At the instant when the input frequency step is applied, $t = 0_+$, the frequency error jumps momentarily to $\Delta\omega$ – the value of the input frequency step – though at this time the phase error is still zero. The frequency error then becomes smaller whilst the phase error becomes greater until, at $\omega_n t = 1$, the frequency error is momentarily zero and the phase error peaks at a value of ($\Delta\omega/e\omega_n$). In fact, at any crossing of the frequency axis, the phase error must be either at a maximum or minimum value because frequency is the time derivative of phase. Beyond $\omega_n t = 1$, the phase error steadily declines to zero whilst the frequency error reaches a peak negative value and then declines to zero. It is evident from the closed contour that the conditions a long time after the application of the frequency step are the same as those beforehand and thus there have been no cycle slips during the reacquisition process.

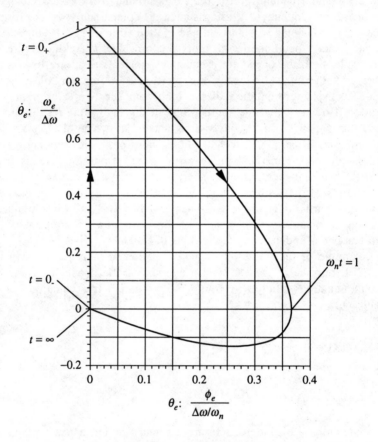

Figure 3.8 Phase-plane plot of a second order type II loop subject to a $\Delta\omega$
frequency step at $t = 0$

By a similar method, a phase-plane plot may be produced for the case of a phase step applied to the same loop. The frequency error response, for this case, may be found by differentiating the phase error response previously derived in equation (3.10),

$$\phi_e(\omega_n t) = \Delta\phi(1 - \omega_n t)e^{-\omega_n t} \qquad (3.19a)$$

and

$$\omega_e(\omega_n t) = \frac{d\phi_e(\omega_n t)}{dt} = \omega_n \Delta\phi(\omega_n t - 2)e^{-\omega_n t} \tag{3.19b}$$

these two expressions forming a pair of parametric equations describing the required phase-plane plot. Before producing such a plot, it is interesting to note the form of the frequency transient in the time domain, as shown in figure 3.9.

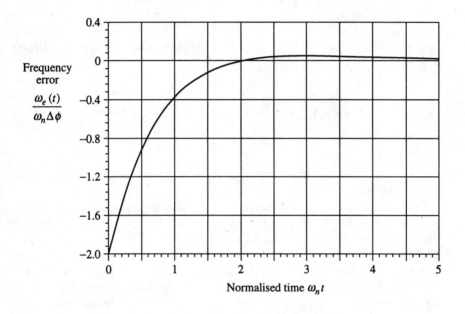

Frequency error $\dfrac{\omega_e(t)}{\omega_n \Delta\phi}$

Normalised time $\omega_n t$

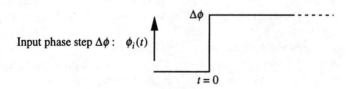

Input phase step $\Delta\phi$: $\phi_i(t)$

$t = 0$

Figure 3.9 Frequency error transient of a second order type II PLL ($\zeta = 1$) resulting from a step change in input phase

A phase step of $\Delta\phi$ is seen to produce an initial frequency error of $\omega_e(0) = -2\omega_n\Delta\phi$ which eventually decays to zero, passing through zero at $\omega_n t = 2$. The sudden frequency error at $t = 0$ may be reconciled by considering the behaviour of the loop components on the introduction of a phase step. If the input phase of a previously settled loop suddenly increases by $\Delta\phi$ then there is an immediate change in the phase detector output voltage of $K_p\Delta\phi$. This, in turn, leads to an instantaneous change in the loop filter output voltage of $(R_2/R_1)K_p\Delta\phi$ and therefore a change in the fed-back VCO frequency of $(R_2/R_1)K_pK_v\Delta\phi/N$. From the loop design equations:

$$\omega_n = \sqrt{\frac{K_pK_v}{NR_1C}} \text{ and } \zeta = \frac{R_2}{2}\sqrt{\frac{K_pK_vC}{NR_1}} = \frac{\omega_nR_2C}{2}$$

the instantaneous frequency error at $t = 0$, being defined as $\omega_e(0) = \omega_i(0) - \omega_o(0)/N$, is thus

$$\omega_e(0) = -\frac{R_2}{R_1}\frac{K_pK_v}{N}\Delta\phi \equiv -\frac{R_2}{R_1}\omega_n^2R_1C\Delta\phi \equiv -R_2C\omega_n^2\Delta\phi \equiv -2\zeta\omega_n\Delta\phi$$

$$(3.20)$$

and this confirms the frequency error of $-2\omega_n\Delta\phi$ at $t = 0$ in a loop with unity damping factor.

The phase-plane plot for the case of a phase step applied to the same second order type II loop, based on equation (3.19), is shown in figure 3.10. As before, the loop is initially in a stable equilibrium so there is no phase or frequency error. When the phase step is applied, at $t = 0_+$, the phase error momentarily becomes $\Delta\phi$ and the frequency error becomes $-2\omega_n\Delta\phi$. This is in contrast to the case of an input frequency step where only a frequency error is experienced at $t = 0_+$. Both phase and frequency errors now diminish, the phase error passing through zero when $\omega_nt = 1$ and the frequency error passing through zero (and simultaneously the phase error passing through a minimum) when $\omega_nt = 2$. Beyond this time the frequency and phase errors rapidly diminish and the system again returns to equilibrium with no cycle slips.

A more challenging situation is the case of a loop experiencing a step change in input frequency which is sufficiently large to take the phase detector beyond its linear operating range, thus introducing the possibility of cycle slips. We shall now consider such an example based on a sequential phase detector with a linear range of $\pm\pi$ radians (such as an RS flip-flop phase detector) subject to a frequency step which would produce a peak phase error of 1.05π radians.

From equation (3.13), the phase error transient arising from a frequency step of $\Delta\omega$ is

$$\phi_e(t) = \frac{\Delta\omega}{\omega_n}\omega_nte^{-\omega_nt}$$

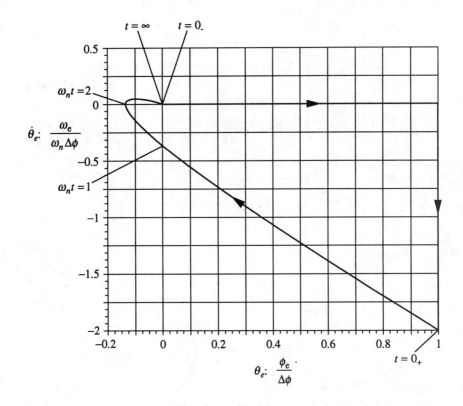

Figure 3.10 Phase-plane plot of a second order type II loop subject to a $\Delta\phi$ phase step at $t = 0$

which has a peak value of $(\Delta\omega/e\omega_n)$ at $\omega_n t = 1$. For a peak phase error of 1.05π radians the required frequency step is therefore $\Delta\omega = 1.05\pi e\omega_n$ and this defines the form of the initial phase transient plotted in figure 3.11. In this situation, the phase error transient exceeds the $\pm\pi$ rads phase detector operating range at time $\omega_n\tau = 0.719$ and at this point in time the phase detector output swings from one extreme, representing $+\pi$ rads phase error, to the opposite extreme, representing $-\pi$ rads phase error. As far as the loop is concerned, it appears as if there is a -2π rads step in the input phase and so a convenient method of analysing the situation is by introducing a -2π rads step at $\omega_n\tau = 0.719$ resulting in a second transient *in addition* to the first transient due to the initial frequency step. This approach relies on the linear properties of Laplace transforms in which the response to a number of separate excitations is equal to the sum of the

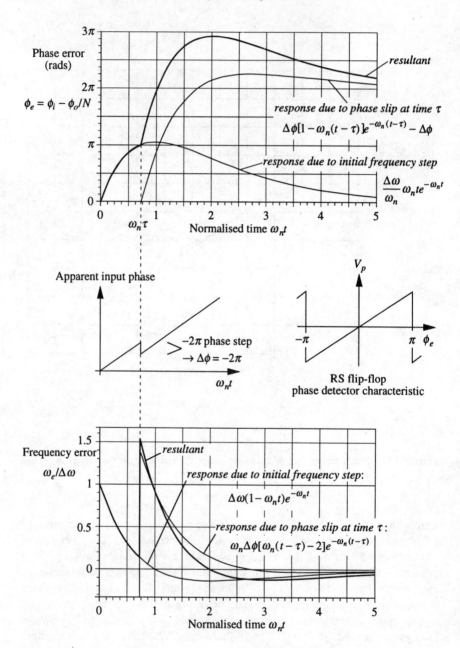

Figure 3.11 Response of a second order type II PLL, $\zeta = 1$, with a modulo-π phase detector to a frequency step, $\Delta\omega = 1.05\pi e\omega_n$, producing a phase transient beyond the linear limit of the phase detector

responses to each individual excitation. So, from time $t = \tau$, a second phase transient is formed, as described by the following (from equation (3.10)):

$$\phi_e(t) = \Delta\phi\big[1 - \omega_n(t-\tau)\big]e^{-\omega_n(t-\tau)} - \Delta\phi$$

where $\Delta\phi = -2\pi$ and the additional $-\Delta\phi$ term allows for the fact that the input phase step is apparent and not actually real.

The resultant phase transient is the sum of these two individual transients which, in this particular situation, is given by

$$\phi_e(t) = 1.05\pi e \omega_n t e^{-\omega_n t} \quad \text{for } \omega_n t < 0.719$$

$$= 1.05\pi e \omega_n t e^{-\omega_n t} - 2\pi\big[1 - (\omega_n t - 0.719)\big]e^{(-\omega_n t + 0.719)} + 2\pi \qquad (3.21)$$

$$\text{for } \omega_n t \geq 0.719$$

and this is plotted in figure 3.11. It is evident that the phase transient initially follows a closed contour – which would return to zero phase error – but then, when the phase detector limit is exceeded, changes to a new path which eventually settles at a phase error of 2π rads. This, of course, means that there has been a single cycle slip. The peak phase error is 2.93π rads which indicates that the loop comes within 0.07π rads of slipping a further cycle.

To produce a phase-plane plot, the frequency error transient must also be obtained as the sum of the transients resulting from the initial frequency step and an apparent phase step at time $t = \tau$,

$$\omega_e(t) = \Delta\omega(1 - \omega_n t)e^{-\omega_n t} \quad \text{for } t < \tau$$

$$= \Delta\omega(1 - \omega_n t)e^{-\omega_n t} + \omega_n\Delta\phi\big[\omega_n(t-\tau) - 2\big]e^{-\omega_n(t-\tau)} \quad \text{for } t > \tau$$

Substituting the frequency step, $\Delta\omega = 1.05\pi e \omega_n$, the apparent phase step, $\Delta\phi = -2\pi$ and $\omega_n\tau = 0.719$ gives the resultant frequency transient,

$$\omega_e(t) = \Delta\omega(1 - \omega_n t)e^{-\omega_n t} \quad \text{for } \omega_n t < 0.719$$

$$= \Delta\omega(1 - \omega_n t)e^{-\omega_n t} - \frac{2\Delta\omega}{1.05e}\big[\omega_n t - 2.719\big]e^{(0.719 - \omega_n t)} \quad \text{for } \omega_n t \geq 0.719$$

$$(3.22)$$

This is also plotted in figure 3.11 and reveals that the effect of the phase slip at time $t = \tau$ is to add a frequency transient some 40% larger than the original frequency step, $\Delta\omega$.

Equations (3.21) and (3.22) may now be combined to produce the phase-plane plot, as shown in figure 3.12. At time $t = 0_-$, immediately before the frequency step is applied, both frequency and phase errors are zero, indicating the case of a settled, locked loop. At the instant when the input frequency step is applied, $t = 0_+$, the frequency error jumps momentarily to $\Delta\omega$ though at this time the phase error is still zero. The plot then begins to follow the contour for a linear loop operating without any cycle slips (as given in figure 3.8) with the frequency error diminishing as the phase error grows. However, at time $t = \tau$ when the phase error reaches the limit of the phase detector operating range, π rads, the nature of the transient suddenly changes with the frequency error jumping to $1.54\Delta\omega$ as the phase error accelerates towards a final settling value of 2π rads.

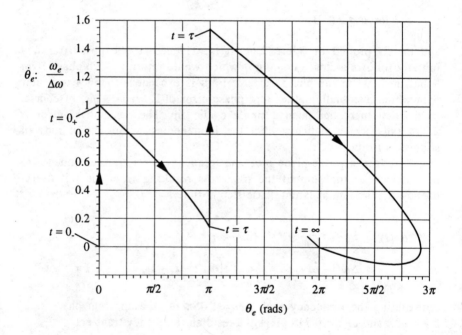

Figure 3.12 Phase-plane plot representing the non-linear loop action depicted in figure 3.11

It is interesting to note that the frequency error when the phase crosses 2π rads is slightly lower (around 12%) than the initial frequency error. This is the reason why the second stage of the transient results in a smaller peak phase error with respect to the final, settled, value and is the basis of acquisition in PLLs in general. If the initial frequency error had been much greater than $1.05\pi e\omega_n$ then many cycle slips might have occurred, though the system would have still eventually converged to a state of stable equilibrium. At this stage the value of phase-plane plots in representing transient, non-linear behaviour should be appreciated. The frequency/phase trajectory of such plots is certainly a helpful way of visualising effects such as cycle slips.

Now is a good time to consider some of the basic properties and definitions relating to phase-plane plots. From figures 3.8, 3.10 and 3.12 it would appear that phase-plane contours may only follow a clockwise path – and this is indeed the general case. There are periodic equilibrium points, or singularities, at zero frequency error and at a phase error sufficient to give zero phase detector output. For analogue multiplier phase detectors, for example, these singularities occur at $\theta_e = \pi/2 + n\pi$ and alternate between points of stable and unstable equilibrium, corresponding to negative and positive loop gain, respectively. A point of unstable equilibrium is termed a saddle point where, of course, operation cannot be sustained for any length of time. Also, as previously mentioned, the phase-plane plot always crosses the phase axis at right-angles.

A feature of type II PLLs (and, indeed, type II control loops in general) is that the instantaneous values of θ and $\dot{\theta}$ at any given time fully define the ensuing phase and frequency transients. In other words, any point on the phase-plane provides sufficient information (given that the loop properties, ω_n and ζ, are known) to generate the complete phase-plane plot and therefore predict the future form of the phase and frequency transients. Similarly, for a type I loop, the instantaneous phase error alone is sufficient for this purpose.

To demonstrate this fact, we can look at the now familiar case of a frequency step, $\Delta\omega$, applied to a second order type II loop at time $t = 0$. The general form of the ensuing phase and frequency transients, as previously derived, are shown in figure 3.13. Now, suppose at some time after the initial frequency step, t_1 say, the instantaneous phase and frequency errors, ϕ'_e and ω'_e, are observed (measured, in practice). Then, without any knowledge of the original excitation, we are asserting that it is possible to exactly determine the future behaviour of the transients. It is possible to show that this is true by finding the necessary combination of phase and frequency steps which, if applied at time $t = t_1$, would give rise to the phase and frequency errors, ϕ'_e and ω'_e. Referring to figure 3.13, the necessary phase and frequency steps are

$$\Delta\phi' = \frac{\Delta\omega}{\omega_n}\omega_n t_1 e^{-\omega_n t_1} \tag{3.23}$$

and

$$\Delta\omega' = \Delta\omega(1 - \omega_n t_1)e^{-\omega_n t_1} + 2\omega_n\Delta\phi' \equiv \Delta\omega(1 + \omega_n t_1)e^{-\omega_n t_1} \qquad (3.24)$$

where the $2\omega_n\Delta\phi'$ term – and this is quite a subtle point – is necessary to cancel the $-2\omega_n\Delta\phi'$ frequency shift resulting from the phase step (previously observed and shown in figure 3.10).

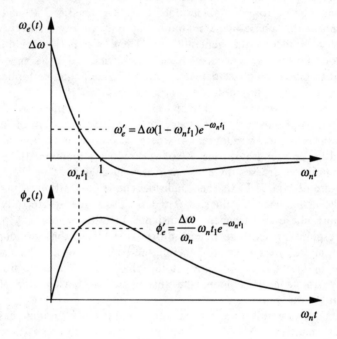

Figure 3.13 Construction used to demonstrate that any point in the phase-plane defines the ensuing phase and frequency transients

Having found the phase and frequency steps which will provide the correct instantaneous phase and frequency error values at time $t = t_1$, it remains to confirm that the resulting transient is indeed identical to that which would arise from the initial $\Delta\omega$ frequency step at time $t = 0$. To do this we derive the phase and frequency errors resulting from simultaneous phase and frequency steps of $\Delta\phi'$ and $\Delta\omega'$ applied at time $t = t_1$. Starting with the frequency error:

$$\omega_e(t) = \Delta\omega'\big[1-\omega_n(t-t_1)\big]e^{-\omega_n(t-t_1)} + \Delta\phi'\omega_n\big[\omega_n(t-t_1)-2\big]e^{-\omega_n(t-t_1)}$$

where the first term represents the frequency transient arising from a *frequency* step of $\Delta\omega'$ and the second term represents the frequency transient arising from a *phase* step of $\Delta\phi'$. Substituting these phase and frequency steps from equations (3.23) and (3.24),

$$\omega_e(t) = \Delta\omega(1+\omega_n t_1)\big[1-\omega_n(t-t_1)\big]e^{-\omega_n t} + \Delta\omega(\omega_n t_1)\big[\omega_n(t-t_1)-2\big]e^{-\omega_n t}$$

$$\equiv \Delta\omega e^{-\omega_n t}\big[1+\omega_n t_1-\omega_n(t-t_1)-\omega_n^2 t_1(t-t_1)+\omega_n^2 t_1(t-t_1)-2\omega_n t_1\big]$$

which reduces to

$$\omega_e(t) = \Delta\omega(1-\omega_n t)e^{-\omega_n t} \tag{3.25}$$

which is, of course, identical to the transient resulting from a frequency step of $\Delta\omega$ applied at time $t = 0$. This result demonstrates, albeit for a specific example, that the ensuing frequency error transient is defined by *any* point in the phase-plane characteristic. The same result is obtainable for the phase error transient – though the proof of this is left as an exercise for the reader!

4 Modulation

4.1 Single-point modulation

An important use of PLLs is in the phase or frequency modulation of a stable carrier wave or, conversely, in the demodulation of phase or frequency modulated signals. PLLs are attractive in modulator applications due to the combination of controllable modulation with a highly stable and adjustable carrier frequency; they are also attractive in certain demodulator applications because of their superior performance at low S/N ratios. The basic arrangement of a PLL modulator is shown in figure 4.1. Angle modulation is easily implemented by adding a modulating signal either before the loop filter for phase modulation or after the loop filter for frequency modulation.

Phase modulation is achieved by effectively adding an offset, V_{pm}, to the phase detector output voltage. The loop responds by adjusting the VCO phase by an amount $\Delta\phi_o$ so that the phase detector output voltage, $K_p\Delta\phi_o/N$, opposes this offset and a new equilibrium point is reached. Thus the steady-state phase deviation is $\Delta\phi_o = NV_{pm}/K_p$. Because the loop is modulated before the loop filter it is less able to respond to modulation frequencies beyond the loop natural frequency and so the modulator exhibits a low-pass characteristic.

In the case of frequency modulation, the addition of a signal after the loop filter acts to directly modulate the VCO frequency provided that the loop response is sufficiently slow to prevent the loop filter output from opposing the modulating signal. In other words, the loop maintains a constant *average* frequency whilst allowing rapid modulation around this frequency. This gives rise to a high-pass characteristic with a peak frequency deviation of $\Delta\omega_o = V_{fm}K_v$ at high modulation frequencies.

In both cases it is important that the phase detector operating range is not exceeded. For a phase/frequency detector this limits the peak phase deviation of the VCO to $2\pi N$ radians, and hence the modulation index is limited to $2\pi N$. In virtually all frequency synthesiser applications, this constraint is insignificant because of the high division ratio.

The loop response to a sinusoidal modulating signal component of frequency ω can be derived from the signal flow graph of figure 4.1. For phase modulation, the loop transfer function is

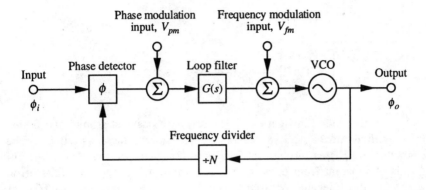

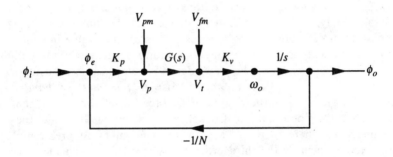

Figure 4.1 PLL with modulation capability and corresponding signal flow graph

$$\frac{\phi_o(s)}{NV_{pm}(s)} = \frac{\dfrac{K_v G(s)}{Ns}}{1 + \dfrac{K_p K_v G(s)}{Ns}} = \frac{1}{K_p}\frac{\phi_o(s)}{N\phi_i(s)} = \frac{2\zeta\omega_n s + \omega_n^2}{K_p(s^2 + 2\zeta\omega_n s + \omega_n^2)}$$

$$(4.1)$$

This is identical in shape to the closed-loop transfer function, $\phi_o(s)/N\phi_i(s)$, and represents a second order low-pass characteristic with a roll-off proportional to $1/s$ at high frequencies, giving a slope of 20 dB/decade. The magnitude and phase of the response are given by

$$\left|\frac{\phi_o K_p}{NV_{pm}}\right|^2 = \frac{1 + 4\zeta^2(\omega/\omega_n)^2}{\left[1 - (\omega/\omega_n)^2\right]^2 + 4\zeta^2(\omega/\omega_n)^2}$$

$$(4.2a)$$

and

$$\arg\left[\frac{\phi_0 K_p}{NV_{pm}}\right] = \tan^{-1}\left[\frac{-2\zeta(\omega/\omega_n)^3}{1+(4\zeta^2-1)(\omega/\omega_n)^2}\right] \qquad (4.2b)$$

and these are plotted in figure 4.2. The response peaks at around $\omega = 0.8\omega_n$, the magnitude of the peak increasing as the damping factor is reduced. The characteristic passes through 0 dB at $\omega = \sqrt{2}\omega_n$ irrespective of the value of ζ (this is apparent from inspection of equation (4.2a)) and then falls off at a rate of 20 dB/decade. At high frequencies the response tends to $2\zeta\omega_n/\omega$, so there is an advantage in using a lower damping factor although this is countered by a corresponding increase in peak magnitude. The general low-pass characteristic is useful to suppress phase noise beyond $\pm2\omega_n$, approximately, from the carrier frequency of the input signal. This is important in demodulator applications, but is of little value in synthesisers or other applications where the input S/N ratio is guaranteed to be high.

The AC component of the phase detector output can be considered as a source of phase modulation – giving rise to *reference sidebands* – and figure 4.2(a) indicates the degree of suppression of this component. It is clear that the loop natural frequency must be well below the input frequency for there to be an appreciable amount of suppression. Higher order loops, such as the third order loop considered in chapter 7, have a steeper roll-off and are capable of greater suppression of such unwanted modulation components. Example 3 considers the relation between loop parameters and reference sidebands for a loop containing an analogue multiplier phase detector.

The phase response of the PM characteristic, as described by equation (4.2b), is shown in figure 4.2(b). This indicates a progressively increasing phase lag which approaches 90° when the modulating frequency is very much higher than the loop natural frequency. The phase response is only occasionally of interest in PLL modulation applications but is of importance when designing a nested loop (i.e. a loop within a loop). In such an application it is important that the outer loop has a lower natural frequency than the inner loop, otherwise the phase delay associated with the inner loop will cause a degradation in phase margin and may result in instability. In order to avoid a phase margin degradation of more than 10°, inspection of figure 4.2(b) indicates that the inner loop should have a natural frequency of no more than half that of the outer loop.

From the signal flow graph of figure 4.1, the loop transfer function for a frequency modulation signal, V_{fm}, added to the loop filter output is given by the following,

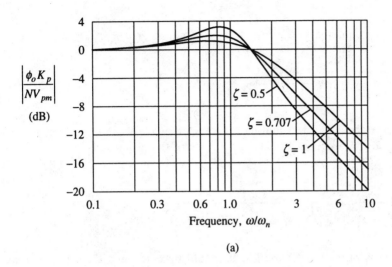

(a)

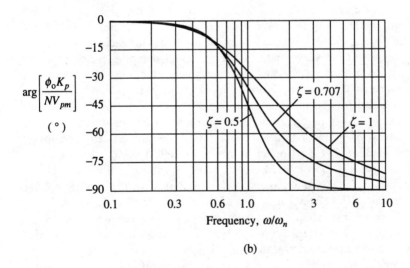

(b)

Figure 4.2 Phase modulation characteristics of a second order type II PLL;
(a) amplitude response and (b) phase response

$$\frac{\omega_o(s)}{V_{fm}(s)} = \frac{K_v}{1 + \dfrac{K_p K_v G(s)}{Ns}}$$

and, using equation (3.6), this can be written,

$$\frac{\omega_o(s)}{V_{fm}(s)} = \frac{K_v s^2}{s^2 + 2\zeta\omega_n s + \omega_n^2} \tag{4.3}$$

This is a second order high-pass characteristic with a roll-off of 40 dB/decade at low frequencies. The magnitude of the response is

$$\left|\frac{\omega_o}{K_v V_{fm}}\right|^2 = \frac{(\omega/\omega_n)^4}{\left[1 - (\omega/\omega_n)^2\right]^2 + 4\zeta^2 (\omega/\omega_n)^2} \tag{4.4}$$

and this is plotted in figure 4.3. It should be noted that the y-axis scaling is different to figure 4.2(a) and the roll-off rate is, in fact, twice that of the phase modulation characteristic. The 3 dB cut-off frequency is equal to the loop natural frequency for a damping factor of 0.707, which is in contrast to the case of a phase modulator where the 3 dB cut-off frequency is twice the loop natural frequency. At damping factors well below 0.707 the response has a noticeable peak, whereas for damping factors well above 0.707 the roll-off is very gradual. The optimum damping factor of 0.707 gives rise to a Butterworth high-pass response producing the maximally flat characteristic seen in figure 4.3. This is quite a good choice of damping factor for FM modulators.

An example of a practical implementation of an FM modulator is given in the circuit diagram of figure 4.4. This is a low power VHF transmitter based around a ÷8/9 dual-modulus prescaler and a CMOS PLL chip. A third order type II loop is used (see chapter 7) with an additional op-amp summer stage to add the desired FM modulation at the tuning voltage input of the VCO. The loop natural frequency is 50 Hz in order to accommodate a modulation frequency range varying from around 30 Hz and upwards – consistent with a VHF FM broadcast station or a high quality audio transmitter. Because of the low loop natural frequency, the circuit tunes very slowly – at a speed of around 20 MHz/s. A modified Colpitts oscillator, based on a BFY90 transistor, is used as the VCO with tuning obtained from a pair of varactor diodes. Although it is possible to construct an RC oscillator at VHF frequencies, the LC design of figure 4.4 offers significantly better spectral purity. The VCO output is buffered by an additional transistor amplifier to prevent any variations in the output load from affecting VCO performance. The MC145152 PLL chip has an additional lock detect output which, in this circuit, controls a pair of PIN diodes in order to disable the RF output when the loop is unlocked. This prevents the radiation of any unwanted frequencies and any undue interference which might otherwise be caused by switching between frequencies.

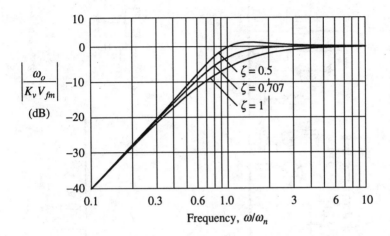

Figure 4.3 Frequency modulation characteristic of a second order type II PLL

In many applications, such as frequency synthesisers, the VCO in a PLL has a much higher residual phase noise level than the reference (input) signal, since the latter is usually a highly stable signal derived, perhaps, from a crystal oscillator. Under these circumstances, any phase noise present on the loop output derives entirely from the VCO noise. However, the loop acts to reduce the residual VCO phase noise present in a free-running VCO, $\phi_n(s)$ say, by locking any phase perturbations to the stable phase reference. The suppression obtained can be modelled by adding phase noise, $\phi_n(s)$, to the output node of the signal flow graph of figure 4.1, resulting in

$$\frac{\phi_0(s)}{\phi_n(s)} = \frac{1}{1+\dfrac{K_p K_v G(s)}{Ns}} = \frac{s^2}{s^2 + 2\zeta\omega_n s + \omega_n^2} \tag{4.5}$$

This is of exactly the same form as the FM characteristic, with the result as shown in figure 4.3. Phase noise is suppressed by 40 dB for every

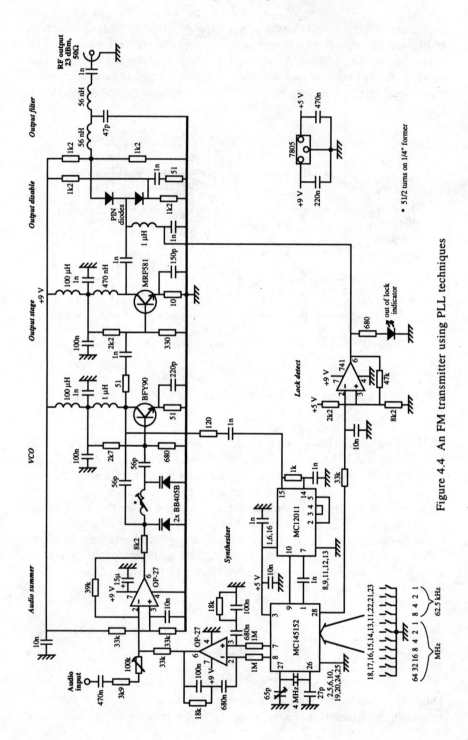

Figure 4.4　An FM transmitter using PLL techniques

decade below the loop natural frequency. Thus, a distinct advantage of using a high loop natural frequency is that a considerable improvement in VCO spectral purity can be achieved. In fact, it is possible to accurately predict the reduction in VCO phase noise due to this mechanism. Referring to figure 4.5, a free-running or unlocked VCO has a certain phase noise spectral density profile which may be easily measured or (not so easily) predicted from the circuit design and component specifications. The loop parameters – natural frequency and damping factor – will probably already have been set by other circuit requirements and so the shape of the PLL FM response will be accurately known. Thus, from our earlier discussion, the phase noise spectral density profile of the locked VCO can be found by multiplying the unlocked VCO phase noise profile by the loop FM response.

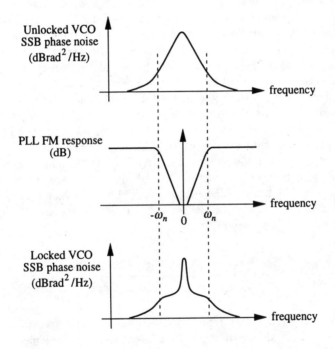

Figure 4.5 Illustration of the improvement in VCO phase noise in a high speed PLL

This results in the suppression of phase noise within approximately $\pm\omega_n$ of the carrier frequency with a progressive increase in suppression at frequencies closer to the carrier. The associated phase noise spectrum is closely mirrored by the power spectrum measurable with a spectrum analyser (drawing on the properties of narrow-band phase modulation outlined in appendix B) and so a characteristic similar to the locked VCO phase noise profile of figure 4.5 would be expected. This general spectral shape is very common in PLL frequency synthesisers and allows an easy approximate estimate of the loop natural frequency. In a communications application, if it is possible to use a reasonably wide loop bandwidth then a considerable improvement in the close-in phase noise (and hence baseband S/N ratio) is clearly obtainable.

4.2 Two-point modulation

The modulation techniques described so far enable the loop to be phase modulated by baseband signals of frequency below $2\omega_n$, approximately, and/or frequency modulated by baseband signals above ω_n, approximately. If, for particular design reasons, the loop natural frequency is too low in the former case or too high in the latter case then it may not be possible to modulate the loop over the desired baseband frequency range. This problem may be overcome to a certain extent by using a compensating filter in the baseband circuitry to extend the pass band. Figure 4.6 presents an example of a compensating filter designed to extend the range of a frequency modulator by a factor of ten. Since the FM response is a high-pass characteristic with a slope of -40 dB/decade then any compensating filter must provide a low frequency lift of 40 dB/decade with a cut-off frequency which closely matches that of the loop. The required filter design can easily be established by considering the factor necessary to translate the loop FM response from a natural frequency of ω_n to an apparent natural frequency of ω_o,

$$\frac{\omega_o(s)}{V_{fm}(s)} = K\frac{s^2 + 2\zeta\omega_n s + \omega_n^2}{s^2 + 2\zeta\omega_o s + \omega_o^2}\cdot\frac{K_v s^2}{s^2 + 2\zeta\omega_n s + \omega_n^2} \qquad (4.6)$$

where the left-hand factor represents the desired filter transfer function. The filter circuit shown in figure 4.6 is a suitable implementation for damping factors of at least 1 and, as an example, for a damping factor of 1 the required filter response is

$$H(s) = K\left[\frac{s + \omega_n}{s + \omega_o}\right]^2$$

and comparing this with the circuit response enables the filter design to be very easily established;

$$H(s) = K\left[\frac{s+\omega_n}{s+\omega_o}\right]^2 = \left[\frac{R_2}{R_1}\right]^2\left[\frac{s+1/(C_2R_2)}{s+1/(C_1R_1)}\right]^2$$

$$\Rightarrow \omega_n = \frac{1}{C_2R_2}; \quad \omega_o = \frac{1}{C_1R_1} \quad \text{and} \quad K = \left[\frac{R_2}{R_1}\right]^2 \tag{4.7}$$

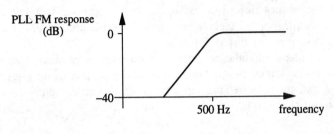

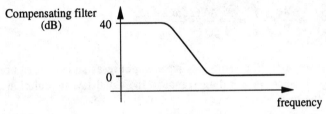

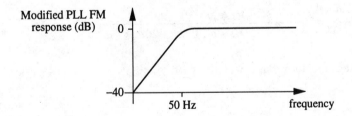

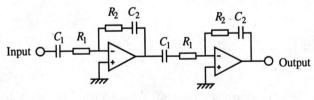

Realisation suitable for $\zeta \geq 1$

Figure 4.6 Extension of FM range by means of a compensating filter

In figure 4.6 an ideal compensating filter is shown which successfully extends the low frequency FM response from 500 Hz down to 50 Hz. However, this approach is not very satisfactory when an increase in bandwidth of more than a decade or so is required because it inevitably leads to high signal levels in the loop filter which gives rise to distortion and noise problems. In addition, it requires close matching of the cut-off frequencies of the filter and loop which is difficult to sustain in frequency synthesiser applications. A better solution, suffering from neither of these limitations, involves a modified combination of the phase and frequency modulation techniques described earlier in this chapter and is known as *two-point modulation*. This technique is equally applicable to PM and FM and will be described for FM as follows.

Firstly, the phase modulation input can be used to produce frequency modulation, as explained in appendix C, by the inclusion of an integrator, of response A/s, in the baseband circuitry – as shown in figure 4.7. This gives, from equation (4.1),

$$\frac{\omega_o(s)}{V_{fm1}(s)} \equiv \frac{A\phi_o(s)}{V_{pm}(s)} = \frac{AN(2\zeta\omega_n s + \omega_n^2)}{K_p(s^2 + 2\zeta\omega_n s + \omega_n^2)} \tag{4.8}$$

This describes exactly the same low-pass response as shown in figure 4.2. By applying the same modulating signal to the usual FM modulation point in the loop, the overall response is

$$\frac{\omega_o(s)}{V_{fm}(s)} = \frac{AN(2\zeta\omega_n s + \omega_n^2)}{K_p(s^2 + 2\zeta\omega_n s + \omega_n^2)} + \frac{K_v s^2}{s^2 + 2\zeta\omega_n s + \omega_n^2} \tag{4.9}$$

Thus, if the integrator has a weighting of

$$A = \frac{K_p K_v}{N}$$

then the combined response is

$$\frac{\omega_o(s)}{V_{fm}(s)} = \frac{K_v(2\zeta\omega_n s + \omega_n^2)}{s^2 + 2\zeta\omega_n s + \omega_n^2} + \frac{K_v s^2}{s^2 + 2\zeta\omega_n s + \omega_n^2} = K_v \tag{4.10}$$

which is clearly constant across all frequencies. The block diagram of a PLL with two-point modulation is shown in figure 4.7. It should be noted that this technique can be used, with the same integrator weighting, for any type of loop filter (see example 8). In synthesiser applications, it is necessary to adjust the weighting to suit the variation in the factor K_v/N as the output frequency changes to avoid a step in the frequency response. A similar technique may be used to achieve two-point phase modulation, requiring the inclusion of a differentiator in the FM input.

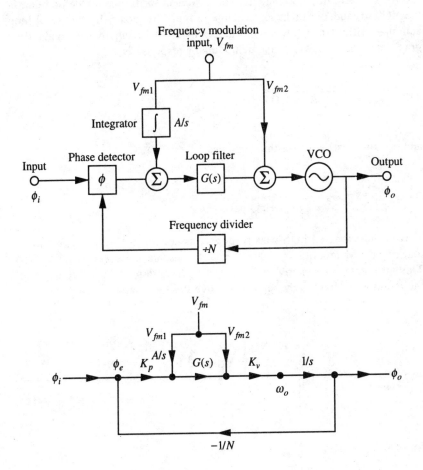

Figure 4.7 PLL with two-point modulation and corresponding signal flow graph

4.3 Demodulation

PLLs may also be used for phase or frequency demodulation. However, as we shall see, their characteristics are opposite to those applying to modulation. Demodulation is achieved simply by applying the modulated signal to the loop input and tapping either the phase detector output voltage for phase demodulation or the loop filter output voltage for frequency demodulation. In the case of phase demodulation, the phase detector output immediately follows any change in input phase while the loop gradually adjusts the VCO phase to track the input phase change – resulting in a high-pass characteristic, whereas for frequency demodulation the VCO frequency tracks any change in input frequency by the usual loop action resulting in a low-pass characteristic. Referring to the signal flow graph of figure 4.8, the phase demodulator response is

$$\frac{V_p(s)}{\phi_i(s)} = \frac{K_p}{1 + \dfrac{K_p K_v G(s)}{Ns}}$$

and, using equation (3.6), this becomes

$$\frac{V_p(s)}{\phi_i(s)} = \frac{K_p s^2}{s^2 + 2\zeta\omega_n s + \omega_n^2} \tag{4.11}$$

This is the expected high-pass response, as shown in figure 4.9, similar to that of an FM modulator where the vertical scale now represents $V_p/K_p\phi_i$. This arrangement is therefore useful as a phase demodulator provided that the modulation frequency band lies above the loop natural frequency.

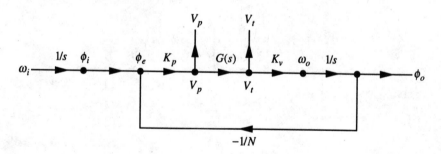

Figure 4.8 Signal flow graph of a PLL used for demodulation

For frequency demodulation the response is

$$\frac{V_t(s)}{\omega_i(s)} = \frac{\dfrac{K_p G(s)}{s}}{1 + \dfrac{K_p K_v G(s)}{Ns}}$$

and, again using equation (3.6), this may be written as

$$\frac{V_t(s)}{\omega_i(s)} = \frac{N(2\zeta\omega_n s + \omega_n^2)}{K_v(s^2 + 2\zeta\omega_n s + \omega_n^2)} \tag{4.12}$$

This is a low-pass response, as shown in figure 4.10, similar to that of a PM modulator where the vertical scale now represents $V_t K_v/N\omega_i$. This arrangement is therefore useful as a frequency demodulator provided that the modulation frequency band lies below approximately twice the loop natural frequency.

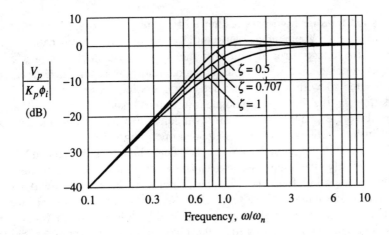

Figure 4.9 Phase demodulation characteristic of a second order type II PLL

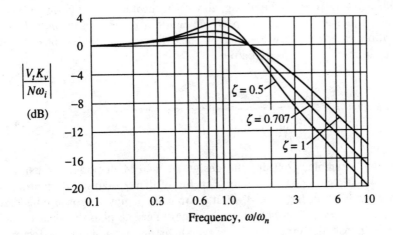

Figure 4.10 Frequency demodulation characteristic of a second order type II PLL

It is important with both PM and FM demodulation that the phase detector is used within its operating range. The phase error between the two applied phase detector inputs is given by

$$\phi_e = \phi_i - \phi_o/N = \phi_i(1 - \phi_o/N\phi_i)$$

and, using equation (3.2), the phase detector error is given by

$$\frac{\phi_e(s)}{\phi_i(s)} = \frac{s^2}{s^2 + 2\zeta\omega_n s + \omega_n^2} \tag{4.13}$$

This is the familiar high-pass response of figures 4.3 and 4.9 showing that, for modulating frequencies substantially beyond the loop natural frequency, the phase error is equal to the input phase deviation − a quite obvious conclusion. So, for a phase demodulator, the input phase deviation is restricted to within the phase detector operating range. This is quite a limitation in the case of analogue multiplier phase detectors, although more elaborate phase detectors are available, such as the tanlock detector (see Gardner, reference 1), with extended operating ranges.

The loop filter design for a frequency demodulator is influenced by two factors. Firstly, the loop natural frequency must be at least one half the highest modulating frequency for the response to be reasonably flat over the modulation bandwidth (figure 4.10). Secondly, on a completely different note, the loop natural frequency should be chosen so that the peak phase detector error, ϕ_e, is within the phase detector operating range and preferably *just* within for optimum noise performance. If the modulation index is less than the phase detector range then this second consideration is unnecessary. For an input signal with sinusoidal modulation of maximum frequency ω_m and of peak deviation $\Delta\omega$, the peak input phase deviation is $\Delta\omega/\omega_m$ (see appendix C) and the corresponding phase error, using equation (4.13), is given by

$$\phi_e(\text{max}) = \frac{\Delta\omega}{\omega_m} \frac{(\omega_m/\omega_n)^2}{\sqrt{\left[1 - (\omega_m/\omega_n)^2\right]^2 + 4\zeta^2(\omega_m/\omega_n)^2}} \tag{4.14}$$

As an example, consider the design of an FM demodulator using an exclusive-OR gate phase detector, capable of demodulating an input signal with a modulation index of 4. Firstly, from frequency response considerations, $\omega_n \geq 0.5\omega_m$. Secondly, the maximum tolerable phase error using this type of phase detector is $\pi/2$ rads and, assuming a damping factor of 0.707, from equation (4.14) the required loop natural frequency is

$\omega_n \geq 1.53\omega_m$. Clearly, the phase error consideration is some three times more important than the frequency response consideration in this instance. Multiplier type phase detectors would normally be used in demodulators because of their superior noise performance. In fact, a properly designed PLL FM demodulator has a lower FM threshold than other types of demodulator, allowing improved performance at low S/N ratios, as illustrated in figure 4.11. The improvement in performance is only slight – around a 2–3 dB threshold extension – but in some demanding applications this may be worthwhile. An example of such an application is satellite communications where link budgets frequently operate on very narrow margins and an extra couple of dBs improvement at the receiver would translate to huge cost savings at the transmitter. However, in other cases where the input S/N ratio is guaranteed to be high, then a digital phase/frequency detector with an operating range of $\pm 2\pi$ rads would be preferred. This caters for modulation indices of up to 2π without any further requirement on the loop natural frequency so that $\omega_n \approx 0.5\omega_m$ could then be used in this example.

In situations where the loop natural frequency lies within the modulation bandwidth, a two-point technique similar to that for modulation could be used. For FM demodulation this entails combining the differentiated phase demodulation output, V_p, with the frequency demodulation output, the differentiator having a weighting of $N/K_p K_v$.

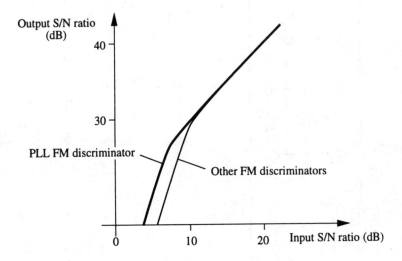

Figure 4.11 Illustration of the FM threshold improvement obtained with PLL FM discriminators

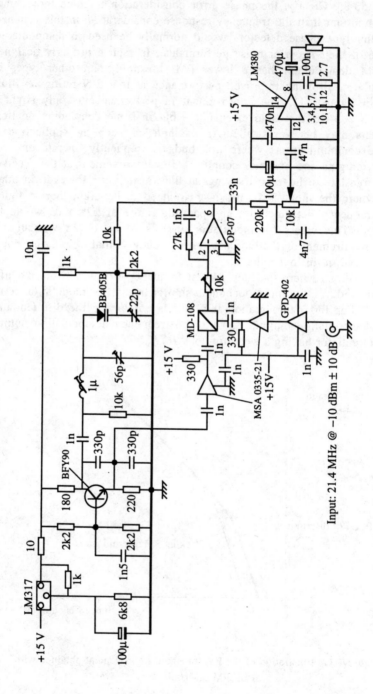

Figure 4.12 Circuit diagram of a PLL FM demodulator

Figure 4.12 shows the circuit diagram of a PLL FM discriminator designed to demodulate an FM signal with a peak frequency deviation of 12 kHz and a maximum modulating frequency of 3 kHz. In contrast to the earlier FM transmitter example, this circuit is based on discrete analogue components. It uses a modified Colpitts oscillator with a deliberately narrow tuning range of 21.4 MHz ± 60 kHz and a high degree of temperature stability – both of which are important to ensure reliable acquisition. The phase detector function is performed by an MD-108 mixer with both inputs buffered by MMIC amplifiers. The loop has a second order type II filter with an adjustable resistor to allow for the possible variation in phase detector gain due to changing input signal levels. The loop natural frequency is 6 kHz with a damping factor of 0.7. From our earlier example (and equation (4.14)) this loop natural frequency may seem a little on the high side, but the design is based on a multiplier phase detector with sinusoidal input signals, in which case the maximum allowable phase detector error is 1 radian. A second order loop is perfectly adequate in this application since the AC phase detector output is at a frequency of 42.8 MHz which is nearly 10000 times the loop natural frequency. In loops without feedback dividers, such as this, any reference sidebands are actually at harmonics of the VCO frequency and are rarely of concern. The audio output is tapped from the VCO tuning input and both high-pass and low-pass filtered with a simple passive first order RC filter having 20 Hz lower and 3.5 kHz higher cut-off frequencies and then applied to an IC audio amplifier. The design is very different from the earlier FM transmitter example in that discrete components are used and loop filter design is not critical; however, VCO stability and phase detector offsets *are* of great importance.

5   Noise Performance

5.1   Noise bandwidth

PLLs can be thought of as tracking filters which admit input signal and phase noise components within approximately $\pm\omega_n$ of the carrier frequency and exclude components outside this range. The effects on the loop output of additive noise at the input of a PLL will be considered in this chapter.

A convenient measure of the ability of a PLL to reject input noise is its *equivalent rectangular noise bandwidth*, often abbreviated to noise bandwidth. A real life filter transfer function, such as that of figure 4.2, can be approximated by a hypothetical transfer function with a perfectly flat passband and an infinitely sharp cut-off, of bandwidth B_L Hz say, where B_L is chosen so that equal noise power is transferred by both filters. B_L is then termed the noise bandwidth of the filter. This definition assumes that the filters are presented with a uniformly distributed noise spectrum, which is usually the case. The situation is illustrated in figure 5.1 where a filter of amplitude response $G(f)$ is presented with input noise of spectral density η W/Hz. The noise power passed by the filter in a small bandwidth df is, of course, $\eta|G(f)|^2 df$ and so the total noise power transferred is

$$noise\ power \;=\; \eta \int_0^\infty \left|G(f)\right|^2 \; df$$

This compares with the noise power that would be allowed through a filter with a perfectly rectangular shape and of cut-off frequency B_L, which is simply ηB_L, and so the one-sided equivalent rectangular noise bandwidth is

$$B_L \;=\; \int_0^\infty \left|G(f)\right|^2 \; df \tag{5.1}$$

The noise bandwidth of any practical filter is almost always greater than the corresponding 3 dB cut-off frequency. In fact, filters with a more gradual roll-off generally have higher noise bandwidths in relation to their 3 dB cut-off frequencies as a result of the additional noise which is passed by the filter beyond its nominal cut-off point. As a simple example, the

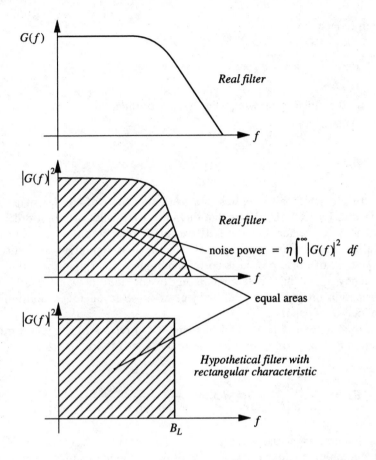

Figure 5.1 Definition of equivalent rectangular noise bandwidth

noise bandwidth of a first order low-pass filter may be considered, with the following transfer function,

$$G(s) = \frac{\omega_o}{s + \omega_o}$$

The noise bandwidth of this filter, from equation (5.1), is thus

$$B_L = \frac{1}{2\pi} \int_0^\infty |G(\omega)|^2 \, d\omega \quad \text{Hz} = \frac{1}{2\pi} \int_0^\infty \frac{\omega_0^2}{\omega^2 + \omega_0^2} \, d\omega \quad \text{Hz}$$

which reduces to

$$B_L = \frac{\omega_o}{2\pi}\left[\tan^{-1}(\omega/\omega_o)\right]_0^\infty \equiv \frac{\omega_o}{4}$$

and so the noise bandwidth, in Hz, as a function of the 3 dB cut-off frequency, f_o, is

$$B_L = \frac{\pi}{2}f_o$$

This result, which also applies to a band-pass filter of similar shape and a first order type I PLL, shows that the noise bandwidth of a first order filter is some 57% greater than the 3 dB cut-off frequency. In a communications situation where noise is an important issue it is important to realise that the amount of noise passed is correspondingly greater than the 3dB cut-off frequency – in this case amounting to an additional 1.96 dB. Some information on the interesting subject of noise in analogue multipliers is provided in appendix E.

In the context of phase-locked loops, $G(f)$ relates to the phase transfer function $\phi_o/N\phi_i$ giving, for a second order type II loop,

$$B_L = \frac{1}{2\pi}\int_0^\infty \left|\frac{\phi_o(\omega)}{N\phi_i(\omega)}\right|^2 d\omega = \frac{1}{2\pi}\int_0^\infty \left|\frac{2j\zeta\omega\omega_n + \omega_n^2}{\omega_n^2 - \omega^2 + 2j\zeta\omega\omega_n}\right|^2 d\omega \quad \text{Hz}$$

This integral is actually quite difficult to evaluate but may be tackled using complex contour integration with the following result,

$$B_L = \frac{\omega_n}{2}\left[\zeta + \frac{1}{4\zeta}\right] \equiv \pi f_n\left[\zeta + \frac{1}{4\zeta}\right] \quad \text{Hz} \tag{5.2}$$

This shows that the noise bandwidth is at least π times the loop natural frequency and therefore, because this is a one-sided bandwidth, input noise is admitted within a bandwidth of at least $2\pi f_n$ centred on the input carrier frequency. By differentiating equation (5.2) it is easy to deduce that minimum noise bandwidth is achieved with a damping factor of 0.5, although the noise bandwidth for unity damping factor is only 25% or 0.97 dB higher. It is now a good time to summarise the properties of the three commonly used damping factors. $\zeta = 0.5$ gives minimum noise bandwidth; $\zeta = 0.707$ gives a Butterworth FM characteristic and $\zeta = 1$ gives critical damping for optimum transient response.

5.2 Phase jitter

Additive noise at the input of a PLL cannot be translated into additive noise at the loop output since the VCO signal may only be phase or frequency modulated by the loop. Instead, input noise close to the carrier frequency is translated into noise components on the DC signals within a locked PLL which results in phase noise on the VCO output signal. This *phase jitter* can readily be quantified by knowing the signal to noise spectral density ratio of the input signal. For the following analysis, the input signal plus band-limited additive noise shall be conveniently represented by

$$V_i(t) = E_c \cos \omega_i t + E_n(t) \cos(\omega_i t + \theta_n(t)) \tag{5.3}$$

where E_c and $E_n(t)$ are the carrier and noise amplitudes, respectively, and $\theta_n(t)$ is the random noise phase. This expression is used to represent an input carrier plus uniformly distributed noise within some arbitrary bandwidth B_i. In practice the input signal would be band-limited to some extent in order to provide a reasonable S/N ratio at the phase detector input although, provided that the input bandwidth is at least twice the loop noise bandwidth, input filtering has no effect on loop phase jitter. Denoting the power spectral density of the input noise by η V^2/Hz, the input noise power is ηB_i and, since the input carrier power is $E_c^2/2$, then the input S/N ratio is

$$S/N_i = \frac{E_c^2}{2\eta B_i} \quad \text{where} \quad \eta B_i \equiv \overline{E_n^2}/2$$

The noise may be resolved into two separate components in phase and in quadrature with the carrier, as shown in figure 5.2. In the context of phase-locked loops, the signal amplitude is irrelevant and it is just the resultant phase of the signal plus noise which is of interest; this assumption is obvious for sequential and exclusive-OR types of multiplier phase detector, though less obvious – but equally correct – for other types of multiplier phase detector. Assuming that the S/N ratio within the loop noise bandwidth is reasonably high, then only the quadrature noise component contributes to the resultant phase noise and the mean square value of the effective phase jitter on the input signal, $\Delta\phi_i(t)$, may be found as follows,

$$\Delta\phi_i = \frac{E_{nq}}{E_c}, \quad \text{so} \quad \overline{\Delta\phi_i^2} = \frac{\overline{E_{nq}^2}}{E_c^2} \quad \text{where} \quad \overline{E_{np}^2} = \overline{E_{nq}^2} = \frac{1}{2}\overline{E_n^2}$$

and expressing this result in terms of the input S/N ratio,

$$\overline{\Delta\phi_i^2} = \frac{\overline{E_n^2}}{2E_c^2} = \frac{\eta B_i}{E_c^2} = \frac{1}{2S/N_i} \tag{5.4}$$

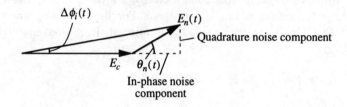

Figure 5.2 Representation of signal plus band-limited noise

Since the input noise is assumed to be uniformly distributed then the phase noise given by equation (5.4) is also uniformly distributed within the input bandwidth B_i. Here the small angle approximation $\tan(\Delta\phi_i(t)) \approx \Delta\phi_i(t)$ has been made which is valid for loop phase jitter values of up to around 12°. The loop responds to the uniformly distributed phase noise apparently present at the input by, effectively, passing all components within $\pm B_L$ of the carrier frequency and rejecting all components outside this range. Assuming $B_L < B_i/2$, the mean square phase jitter on the VCO output is therefore

$$\overline{\Delta\phi_o^2} = \frac{2B_L N^2}{B_i} \overline{\Delta\phi_i^2}$$

and, using equation (5.4), this becomes

$$\overline{\Delta\phi_o^2} = \frac{N^2 B_L}{S/N_i B_i} \tag{5.5}$$

This result can be further simplified by noting that $S/N_i B_i = E_c^2/(2\eta)$ – the input carrier to noise spectral density ratio – which is often denoted by C/n_o, giving

$$rms\ phase\ jitter\ =\ N\sqrt{\frac{B_L}{C/n_o}}\ \ \text{rads} \tag{5.6}$$

This expression is very useful in estimating phase jitter in (amongst others) receiver applications where the input carrier to noise spectral density ratio is derived directly from link budget calculations. It is valid for phase jitters of up to around 12° rms, beyond which the small angle approximation made in the analysis no longer applies and the phase jitter begins to increase more rapidly. It must be appreciated that the phase jitter in loops using sequential phase detectors operating at relatively low input S/N ratios (< 15 dB, say) is very much higher than this result would suggest because the phase detector is not able to average the effects of large noise bursts over a complete input cycle but merely responds to discrete edges of the input waveform. For this reason, multiplier phase detectors *must* be used in applications where the input S/N ratio is low and the preceding analysis may only generally be applied to this case.

Phase jitter calculations are frequently made in communications applications where it is easy to calculate the carrier power to noise spectral density ratio, C/n_o, by standard link budget considerations involving transmit power, transmitter and receiver antenna gains, path losses and receiver noise factor. Figure 5.3 gives an outline of a general receiver involving a PLL. For a given system noise factor, F, the noise power admitted by the pre-PLL filter, of noise bandwidth B, is $KTBF$ and so the noise spectral density (or noise power per unit bandwidth) is $\eta = n_o = KTF$. The incoming carrier power can be estimated for a given communications link, including such factors as fading losses and link margins, and so if the expected input power is C (equivalent to $E_c^2/2$) then the carrier power to noise spectral density ratio is

$$C/n_o\ =\ \frac{C}{KTF} \tag{5.7}$$

As an example of a typical link budget calculation, suppose we have a receiver with a noise figure of 10 dB incorporating a PLL with a noise bandwidth of 100 kHz, which is receiving an incoming signal of level −100 dBm. With the aid of equations (5.6) and (5.7) it is easy to calculate the expected loop phase jitter. Firstly, C/n_o may be conveniently found by manipulating parameters in dB terms (where 0 dBm ≡ 1 mW),

$$C/n_o\ =\ -100\ \text{dBm}\ -10\ \text{dB}\ -10\log_{10}(KT)\ \ \text{dBm/Hz}$$

$$=\ -110\ \text{dBm}\ +174\ \text{dBm/Hz}\ =\ 64\ \text{dBHz}$$

and then the loop phase jitter may be calculated from equation (5.6),

$$rms\ phase\ jitter\ =\ \frac{1}{2}\big[50\ dBHz\ -\ 64\ dBHz\big]$$

$$=\ -7dB\ rads\ \equiv\ 0.2\ rads\ or\ 11.4°\ rms.$$

By analogy with equation (5.4), a quantity known as the loop S/N ratio may be defined,

$$\overline{\Delta\phi_o^2}\ =\ \frac{1}{2S/N_L}$$

which may be expressed as

$$S/N_L\ =\ \frac{S/N_i B_i}{2B_L}\ \equiv\ \frac{C/n_o}{2B_L}\quad\quad\quad\quad\quad(5.8)$$

Unlike phase jitter, loop S/N ratio has no distinct physical meaning because of the absence of an AC signal within a locked loop to compare with any noise; however, it is a useful quantity for specifying other aspects of loop behaviour, such as cycle slipping.

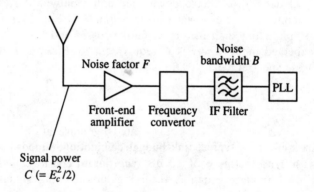

Figure 5.3 A general PLL-based receiver

5.3 Effects of pre phase detector limiting

Signal limiting prior to the loop input is often incorporated in PM or FM demodulator applications to prevent loop behaviour from varying with fluctuations in signal strength and/or to provide a more linear phase detector characteristic. Typically, an analogue multiplier phase detector would be used which would be preceded by a hard limiter and a bandpass filter, as shown in figure 5.4. Alternatively, an exclusive-OR phase detector could be used without the need for a limiter. Although it is not shown in figure 5.4, automatic gain control (AGC) may also be included in a practical design to further reduce the effects of signal fluctuations and has the advantage that it is substantially a linear process.

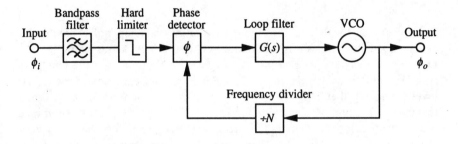

Figure 5.4 PLL with bandpass limited input

The incorporation of limiting has two effects on the signal entering the phase detector. Firstly, the additive S/N ratio is modified by the limiting action. At high input S/N ratios this results in a 3 dB improvement in S/N ratio (since the in-phase noise component now makes no contribution), whereas at low S/N ratios it causes a degradation of approximately 2 dB (as found by Davenport, reference 4). This change in additive S/N, however, does not influence loop phase jitter, as confirmed in both theory and practice, primarily because the phase detector is insensitive to amplitude noise. The second and more important effect is a reduction in signal strength due to the limiter being dominated by noise at low S/N ratios. This directly affects the phase detector gain which is reduced, at low input S/N ratios, by the *limiter signal suppression factor*, courtesy of Davenport (reference 4),

$$\alpha \approx \sqrt{\frac{S/N_i}{4/\pi + S/N_i}} \tag{5.9}$$

From equation (3.3), the loop natural frequency, ω_n, and damping factor, ζ, are proportional to the square root of the phase detector gain and hence the modified values of these parameters, $\omega_{n\alpha}$ and ζ_α, taking into account limiter signal suppression, are

$$\omega_{n\alpha} = \sqrt{\alpha}\omega_n \text{ and } \zeta_\alpha = \sqrt{\alpha}\zeta.$$

Of course, loop noise bandwidth also varies with the limiter signal suppression factor, from equation (5.2), as follows,

$$B_L = \frac{\omega_n}{2}\left[\alpha\zeta + \frac{1}{4\zeta}\right] \tag{5.10}$$

This result shows that at low input S/N ratios the loop noise bandwidth becomes smaller – which is a potentially useful effect. As an example, in a PLL carrier recovery application, it may be sensible to choose the bandpass filter bandwidth such that $\alpha = 0.5$ at the minimum expected input signal level (requiring that $S/N_i = -3.7$ dB). The loop filter may then be designed to provide a damping factor of 0.707 at high input S/N ratios (giving a Butterworth response) which falls to 0.5 at low input S/N ratios (giving minimum noise bandwidth) – the corresponding change in loop noise bandwidth being from $0.530\omega_n$ to $0.354\omega_n$, a 1.8 dB variation. Thus the loop has the desirable properties of Butterworth characteristics at high input signal levels and reduced noise bandwidth at low input signal levels.

Figure 5.5 shows results relating to a highly sensitive pilot carrier receiver designed for use with a communications satellite. The system is designed to receive a pilot carrier having a level as low as –142 dBm with a resulting phase jitter of around 15° rms. The receiver has a noise figure of 2 dB (equivalent to a noise temperature of 475 K when tested using a signal generator at room temperature) and therefore requires a noise bandwidth of around 50 Hz to achieve the phase jitter specification. The chosen design values for the system were a loop natural frequency of 20 Hz, a damping factor of 1.4 at high signal levels and an IF filter bandwidth of 3 kHz giving a limiter signal suppression factor, α, of 0.45 at –142 dBm. The variation of α, noise bandwidth and predicted phase jitter with input signal level may be calculated using equations (5.6) to (5.10) and are tabulated in figure 5.5. It is evident that the design values result in a near two-fold variation in noise bandwidth from low to high signal levels and a change in loop natural frequency from $\sqrt{(0.45)} \times 20$ Hz = 13.4 Hz at –142 dBm to 20 Hz at –120 dBm and above. The variation in damping factor is from 0.94 to 1.4 over a similar signal level range.

The variation of predicted and measured rms phase jitter with input signal level is plotted in figure 5.6. The phase jitter calculation used in this example has been amended slightly to allow for a certain amount of *residual* phase jitter present on the VCO due to oscillator noise, as shown in the following equation,

Signal level (dBm)	C/n_o (T = 475 K) (dBHz)	S/N_p (B_p = 3kHz) (dB)	α	B_L (Hz)	RMS phase jitter (°)		
					Predicted	Measured	
					10 – 10k Hz	10 – 10k Hz	300 – 3k Hz
−143	28.8	−5.9	0.41	47.1	15.5	19.6	6.4
−142	29.8	−4.9	0.45	50.7	14.5	15.0	7.4
−141	30.8	−3.9	0.49	54.4	13.6	13.3	7.7
−140	31.8	−2.9	0.53	58.2	12.8	12.2	7.6
−139	32.8	−1.9	0.58	62.1	12.0	11.6	7.2
−137	34.8	0.1	0.67	69.8	10.6	10.4	7.5
−135	36.8	2.1	0.75	76.9	9.5	10.0	7.4
−133	38.8	4.1	0.82	83.1	8.6	9.7	4.8
−131	40.8	6.1	0.87	87.9	7.8	8.0	6.2
−127	44.8	10.1	0.94	94.2	6.9	7.1	5.8
−117	54.8	20.1	0.99	98.7	6.2	6.2	5.5

Figure 5.5 Measured versus calculated phase jitter values

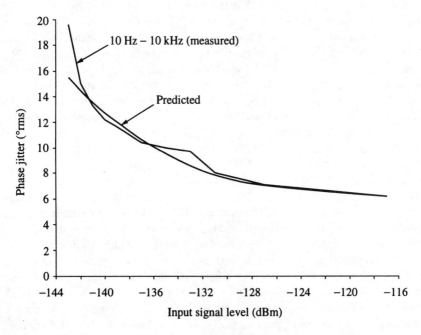

Figure 5.6 Phase jitter variation with input signal strength

$$rms\ phase\ jitter\ =\ \sqrt{N\frac{B_L}{C/n_o}+\overline{\Delta\phi_v^2}}\qquad(5.11)$$

where $\overline{\Delta\phi_v^2}$ is the residual VCO phase jitter which is 6.1° rms (measured) in this example.

The measured and predicted results actually agree very well – owing to the inclusion of limiter signal suppression in the calculations – however the predicted phase jitter does appear to be something of an underestimate at very low input signal levels. This is because the linear assumption on which the phase jitter analysis is based becomes invalid at high phase jitter values and so performance degrades more rapidly than expected.

It is interesting to look at the phase modulation density profile of the system. This requires the use of a phase modulation spectrum analyser which, in its simplest form, consists of a high resolution spectrum analyser with a 6 dB scaling to convert the measured power spectrum to a narrow-band phase modulation spectrum (appendix B). The results of such a measurement made with the pilot carrier receiver are shown in figure 5.7. In figure 5.7(a), at a relatively high input signal level of −120 dBm, the close-in phase noise is quite low and the overall phase noise in the measurement bandwidth of 10 − 10k Hz is noticeably less than that obtained at a somewhat smaller input signal level, −142 dBm, shown in figure 5.7(b). It is quite clear that the phase noise profile at frequencies of 500 Hz and above is almost independent of the input signal level, since this region is well above the loop noise bandwidth and, accordingly, any variation in input signal level results in a change in phase noise restricted largely to within the loop bandwidth. By integrating the phase noise spectral density over the full measurement bandwidth, the total phase jitter may be obtained; indeed this is exactly how the measured results presented in figures 5.5 and 5.6 were arrived at by the automatic network analyser performing the measurement.

An approximate estimate of the integrated phase noise may be made by dividing the plot into a number of frequency bands and adding the mean square phase jitter values in each band. This may be illustrated with reference to the plot of figure 5.7(a) which can be divided into three bands from 10 − 100 Hz, 100 − 1k Hz and 1k − 10k Hz with approximate average phase noise densities of −55 dBrad^2/Hz, −56 dBrad^2/Hz and −60 dBrad^2/Hz, respectively. Within the 10 − 100 Hz band (of width 19.5 dBHz) the integrated phase noise power is −35.5 dBrad^2 or 0.28 mrad^2, within the 100 − 1k Hz band (of width 29.5 dBHz) the integrated phase noise power is −26.5 dBrad^2 or 2.24 mrad^2 and within the 1k − 10k Hz band (of width 39.5 dBHz) the integrated phase noise power is −20.5 dBrad^2 or 8.91 mrad^2. Thus the total integrated phase noise is 0.28 + 2.24 + 8.91 = 11.43 mrad^2 and so the rms phase jitter is 0.107 rads which is equivalent to 6.1° rms.

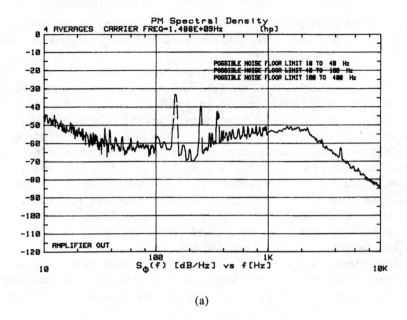

(a)

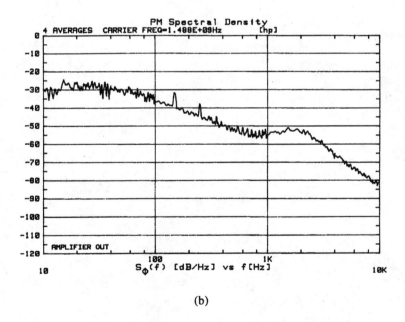

(b)

Figure 5.7 Measured phase modulation density profiles at (a) −120 dBm and (b) −142 dBm input signal levels with loop noise bandwidths of (a) 51 Hz and (b) 97 Hz

This compares very closely with the figure recorded on the automatic analyser (of 6.2° rms) which was computed by a similar method but with many more frequency bands.

It is useful to view the phase detector output waveform on an oscilloscope to get a rough idea of the behaviour of a system in the presence of noise. Some typical waveforms obtained with a PLL based on an analogue multiplier phase detector are shown in figure 5.8. The particular phase detector used in this case is of the type shown in figure 2.2 comprising an analogue switch and an op amp which serves to invert or non-invert an analogue input under the control of a digital signal. At a high input signal level, figure 5.8(a), loop operation is close to perfect with a well limited analogue waveform having sharply defined edges and no evidence of phase noise. At lower levels, figures 5.8(b) and 5.8(c), phase jitter is apparent from the multiple edges visible on the analogue waveform which reflect the variation in VCO phase values. It also appears that the limiting is less effective and there are signs of additive noise in addition to the obvious phase noise. In all three cases the phase detector waveforms indicate that the loop is behaving very well – with no noticeable DC offset or tendency towards loop instability.

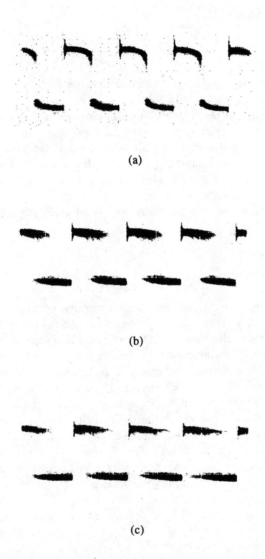

(a)

(b)

(c)

Figure 5.8 Phase detector output waveforms with (a) high (−110 dBm), (b) medium (−134 dBm) and (c) low (−142 dBm) input signal levels

6 Acquisition

The treatment given so far has assumed that the loop is either in lock or close to its stable locked state where the phase error is within the usual phase detector operating range – thus allowing accurate analysis by linear methods. However, there are many occasions when the fed-back VCO frequency is very different to the input frequency and so the loop is far from its stable locked condition; for example, when the loop is initially switched on or as a result of a power supply transient. The process of attaining lock is called *acquisition*, which is a complex subject requiring non-linear analysis and/or simulation, and is of great importance in many PLL circuits.

6.1 Self-acquisition

A PLL which is able to pull into lock naturally without the aid of any additional circuitry is called *self-acquiring*. Any PLL using a phase/frequency detector falls into this category since this type of phase detector provides a very clear indication of the frequency deviation of the loop from its locked condition. Conversely, a PLL which is unable to acquire lock rapidly enough, or at all, by itself may require additional *aided acquisition* circuitry to assist the acquisition process. PLLs using analogue multiplier phase detectors are the most likely to require aided acquisition since their acquisition range (or capture range) is the smallest of all the commonly used phase detectors. The following definitions and results apply to PLLs using analogue multiplier phase detectors and therefore should be regarded as conservative estimates of the performance of PLLs using other types of phase detector.

The *lock-in* range of a PLL is defined as the initial frequency offset range over which the loop is able to acquire lock without any cycle slips occurring. Acquisition then simply results in a phase transient which may be solved analytically for a simple first order type I loop (i.e. containing a passive loop filter). The lock-in limit for a first order type I loop can be shown to be equal to the high frequency loop gain and this may be used as a conservative estimate for higher loop orders so that the lock-in limit is given by

$$\Delta\omega_L \approx \pm\frac{K_p K_v G(\infty)}{N} \tag{6.1}$$

The time taken for the loop to reach phase lock once the frequency error is within the lock-in limit is given approximately by

$$T_L \approx \frac{1}{\omega_n} \tag{6.2}$$

The concept of lock-in range for second order type II loops is somewhat vague since it is a function of both the initial phase and frequency offsets. Nevertheless, the results quoted are useful starting points in any practical design.

If the initial frequency offset is beyond the lock-in limit, ω_L, but within the *pull-in* limit, then the loop is still able to acquire lock but slips cycles during the acquisition process. The acquisition mechanism is now more tenuous, but is quite interesting. Referring to figure 6.1, in isolation, or in an open loop situation, the phase detector forms a sinusoidal beat note equal to the frequency offset of the loop from lock. When the loop is closed, this beat note modulates the VCO frequency and so corrupts the original sinusoidal nature of the beat waveform, introducing a DC component. This DC component then serves to adjust the mean frequency of the VCO and so bring the loop closer to lock. The process then accelerates until lock is acquired. Analysis of the situation is complicated and involves a number of assumptions.

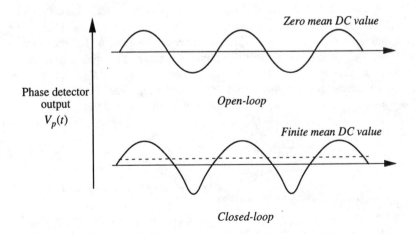

Figure 6.1 Acquisition in a PLL based on a multiplier phase detector

An approximate expression for the pull-in limit, again for the case of a first order loop, is

$$\Delta\omega_p \approx \pm \frac{K_p K_v}{N} \sqrt{2G(0)G(\infty)} \qquad (6.3)$$

Since the DC loop filter gain, $G(0)$, is likely to be much higher than the infinite frequency gain, it is clear that the pull-in limit may significantly exceed the lock-in limit. The corresponding pull-in time is given by

$$T_p \approx \frac{(\Delta\omega)^2}{2\zeta\omega_n^3} \qquad (6.4)$$

showing a $1/\omega_n^3$ dependency, from which it is apparent that for narrow loop bandwidths the pull-in time may be unacceptably long. A further practical factor which has not yet been taken into account is that the pull-in limit may be greatly diminished by the phase detector DC offset. This offset can very easily exceed the small desirable DC component of the phase detector output and, if its polarity is unfavourable, prevent lock from occurring. In a loop using an active filter this invariably means that the loop latches-up at one of its extreme values of VCO frequency. In the author's experience, a good rule-of-thumb estimate for the maximum capture range of a PLL incorporating a multiplier type of phase detector, accounting for the phase detector DC offset, is $\pm 10\omega_n$, though it may be prudent to design any circuit for half of this maximum range to ensure reliable loop operation.

The *hold-in* limit defines the range of frequencies over which the loop is capable of operating and depends only on the DC loop gain. Unlike the other parameters, it is easy to derive a precise expression for the hold-in limit. For an analogue multiplier phase detector, the maximum output voltage is simply equal to the phase detector gain, K_p, and so the maximum frequency range of the VCO (assuming that none of the loop components saturate) is $\pm K_p K_v G(0)$. Thus the hold-in range is

$$\Delta\omega_h = \pm \frac{K_p K_v G(0)}{N} \qquad (6.5)$$

and this is greater than the pull-in range for active loop filters. This result applies with equal accuracy to any design of loop filter.

As an example, consider the case of a PLL using an analogue multiplier phase detector and a second order type II filter, designed for a loop natural frequency of 10 kHz, a damping factor of 0.707 and with the following parameters: $K_p = 1$ V/rad, $K_v = 20$ kHz/V and $N = 1$. From the design formulae, equation (3.3), the filter design requires $R_1C = 31.8$ μs and $R_2C = 22.5$ μs. Thus, the high-frequency filter gain is $G(\infty) = R_2/R_1 = 0.71$ and the DC filter gain, $G(0)$, is equal to the open loop gain of the particular op-amp used in the design – which will be around 100000 (give or take a factor of 10!). From equations (6.1) to (6.4), the lock-in and pull-in limits and times are therefore $\Delta f_L = \pm14.2$ kHz, $T_L = 16$ μs; $\Delta f_p = \pm7.5$ MHz and $T_p = 1.1$ ms from an initial frequency offset of $\Delta f = 100$ kHz. It should be appreciated that the pull-in limit calculated here is strictly a theoretical limit which can only be achieved under perfect conditions. In particular, the presence of any DC phase detector offset will serve to greatly reduce the actual pull-in limit in a practical circuit. The 'rule-of-thumb' estimate of pull-in range in this example is ten times the natural frequency, or ±100 kHz, which is likely to be a much more realistic limit than that calculated above and the associated capture time from an initial frequency offset of 100 kHz is the sum of T_L and T_p which is around 1.1 ms.

Loops incorporating sequential phase detectors are able to acquire lock much more rapidly because of the extended operating range and improved linearity of such phase detectors. Sections 3.3 and 7.3.2 outline the transient behaviour of second and third order type II loops operating within their linear phase detector range and present an accurate analytical method of determining the capture time for given initial conditions. If the conditions, however, are such that the phase error exceeds the phase detector operating range then empirical or numerical methods are again required. As an example of the use of one of the results obtained in section 3.3, equation (3.14) determines the peak phase error resulting from an initial frequency offset, so if this phase error is set to 2π rads for a phase/frequency detector then an expression for the maximum frequency offset allowing operation without cycle slips (namely, the lock-in range) is obtained,

$$\Delta\omega_L = \pm 2\pi e \omega_n$$

This result applies only to a second order type II loop with unity damping factor and with a phase/frequency detector.

The ability of a loop to acquire lock is greatly impaired at low S/N ratios and it is generally considered that the loop S/N ratio must be at least 6 dB in order for acquisition to occur. Indeed, a loop operating at low S/N ratio tends to slip cycles due to the presence of noise. The mean time between these cycle slips can be expressed as a function of the loop S/N ratio. There are several approximate formulae for such cycle slips, the following resulting from work performed by Ascheid and Meyr (reference 5) applies to first order loops, though it is still a useful guide to the performance of second order loops,

$$T_{av} \approx \frac{2}{\omega_n} \exp(\pi S/N_L) \qquad (6.6)$$

This function is plotted in figure 6.2 and shows that the expected time between cycle slips increases very rapidly with the loop S/N ratio. For instance, for a loop S/N ratio of 3 dB the mean time between cycle slips, T_{av}, is around $1000/\omega_n$, while for a small improvement in loop S/N ratio to 10 dB T_{av} increases by some 11 orders of magnitude to around $10^{14}/\omega_n$.

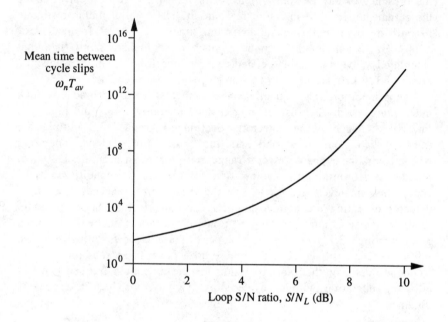

Figure 6.2 Cycle slip variation with loop S/N ratio

6.2 Aided acquisition

Because of the fragility of the acquisition mechanism of loops based on multiplier phase detectors it may not be possible, in certain applications, for the loop to naturally pull into lock. There are several ways of helping a PLL to attain lock. The most common of these involves the use of a DC signal inserted at the loop filter input which linearly sweeps the frequency of the VCO, as shown in figure 6.3. If the sweep rate is sufficiently low, then when lock is imminent the natural action of the loop takes over from the imposed sweep signal, bringing the frequency sweep to a halt and

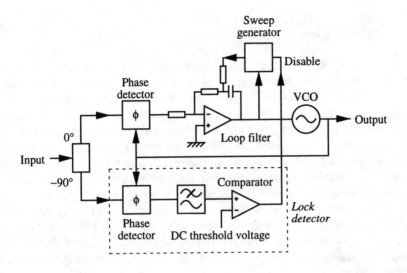

Figure 6.3 PLL with frequency-swept aided acquisition

allowing the loop to lock. A lock detector is then used to disable the sweep signal. Note that the sweep signal is inserted at the junction of the resistor and capacitor feedback components in order to avoid a sudden voltage step on its removal. It is important to know the maximum rate at which the loop may be swept, which can be determined as follows.

Suppose a loop is locked to an input signal which is swept at a uniform rate of $d\omega/dt = K$ rad/s^2. The input phase is the double integral of $d\omega/dt$ which can be represented in Laplace notation,

$$\phi_i(t) = \frac{Kt^2}{2} \quad \text{so} \quad \phi_i(s) = \frac{K}{s^3}$$

Using equation (4.13), the phase error transient resulting from the introduction of a frequency ramp at $t = 0$ is given by

$$\phi_e(s) = \frac{K}{s(s^2 + 2\zeta\omega_n s + \omega_n^2)}$$

and the steady state phase error may be found from the final value theorem,

$$\underset{\lim\, t\to\infty}{\phi_e(t)} \;=\; \underset{\lim\, s\to 0}{s\phi_e(s)} \;=\; \frac{K}{\omega_n^2} \tag{6.7}$$

This result applies to any second order type II PLL, regardless of the damping factor, and shows that the loop is able to track a frequency ramp with a finite phase error. In the case of an analogue multiplier phase detector, the effective operating range is 1 radian (because the response is sinusoidal) which means that the maximum sweep rate that the loop can possibly track is ω_n^2 rad/s^2. This is the absolute limit on the sweep rate that can be used with the circuit of figure 6.3, since beyond this the loop would be unable to arrest the imposed frequency sweep. In fact, a somewhat lower sweep rate is needed if reliable operation is required from the circuit and it has been found that a sweep rate of no greater than $0.5\,\omega_n^2$ allows the method to be used with virtually 100% probability of acquisition occurring on a single sweep. It should be noted that at low loop S/N ratios the sweep rate must be reduced still further to allow reliable operation.

Figure 6.3 also illustrates a method of detecting when phase lock has been acquired. An additional phase detector is used as a *quadrature detector* (i.e. with a 90° phase shift included in one of its inputs) to detect when the loop phase error is 90°, which is the steady state lock condition for a loop incorporating a multiplier phase detector. When the loop is out of frequency lock, the quadrature phase detector output is an AC waveform which is suppressed by the low-pass filter so that the comparator output is low. As the loop enters its lock-in range, the quadrature detector output becomes a DC signal dependent on the loop phase error and with a maximum value when the loop is locked. This DC signal then triggers the comparator, thus providing a clear indication of phase lock. The quadrature detector output also enables the loop to be used as a coherent AM demodulator.

The appearance of a multiplier phase detector output waveform in the presence of a sinusoidal swept-frequency reference signal is shown in the experimental results of figure 6.4. When the sweep direction is positive, i.e. increasing frequency with time (figure 6.4 (a) and (c)), the waveform shifts in such a way as to produce a negative DC component in the phase detector output which, with an integrator-type of loop filter, gives rise to an increasing tuning voltage and therefore an increasing VCO frequency. Conversely, with a negative sweep direction (figure 6.4 (b) and (d)), there is a clear positive DC component in the phase detector output. The phase detector featured in the figure is driven by a digital square-wave signal and a limited analogue signal so that its linear range is virtually ±90°. Thus, applying equation (6.7) to this case where the loop natural frequency is 23 Hz, the maximum possible sweep rate is 5.2 kHz/s which seems

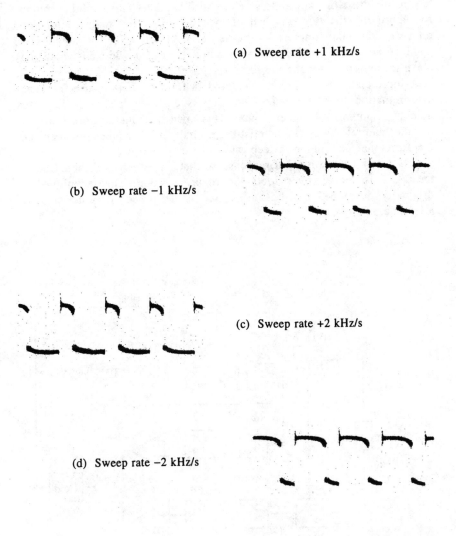

(a) Sweep rate +1 kHz/s

(b) Sweep rate −1 kHz/s

(c) Sweep rate +2 kHz/s

(d) Sweep rate −2 kHz/s

Figure 6.4 Phase detector outputs with a swept reference signal, loop natural frequency = 23 Hz

entirely consistent with the appearance of the waveforms in the presence
of 1 kHz/s and 2 kHz/s frequency sweeps.

An alternative method for aiding acquisition is to initially widen the
loop bandwidth – thus increasing the capture range – and then to narrow
the bandwidth when the loop is close to lock. This is a useful technique
for improving the tuning speed of synthesisers whilst maintaining good
reference sideband suppression, but is not as effective for low signal level
applications since the loop noise bandwidth is also widened which may
make it impossible for the loop to respond.

A more elaborate technique, featured in figure 6.5, is to use frequency
discrimination to detect when the loop is close to lock and then very
quickly removing the sweep signal. This open loop method does not rely
on the natural loop characteristics to halt the frequency sweep and
therefore enables a higher sweep rate to be used, but it does depend on the
discriminator having a rapid response so that the sweep is disabled before
the VCO frequency overshoots the required value. The technique was
originally developed by the author and is described in more detail in
reference 6.

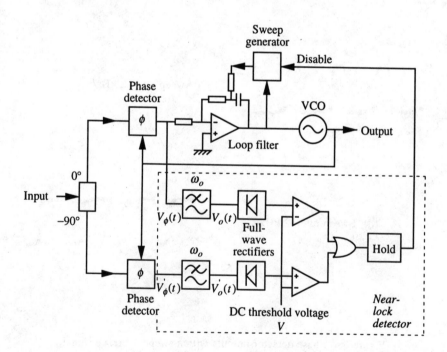

Figure 6.5 Phase-locked loop with rapid frequency sweeping controlled by a
near-lock detector

Referring to figure 6.5, the basis of the technique is a so-called near-lock detector comprising a pair of low-pass filters, rectifiers and threshold detectors, which is used to determine when the loop frequency error is less than approximately the cut-off frequency of the low-pass filters, ω_o. By rapidly disabling the sweep at this point, and by choosing ω_o to be sufficiently low, the loop should be able to pull into lock naturally. However, it is not a trivial matter to determine the cut-off frequency required for a viable design in view of the dynamic nature of the processes involved. The remainder of this chapter is a summary of the numerical analyses leading to a good understanding of the operation of the system, its design and its capabilities.

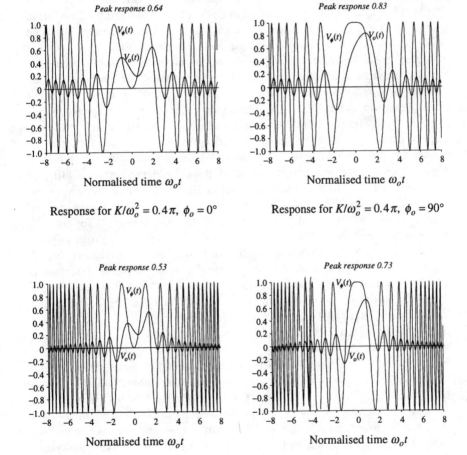

Figure 6.6 Some modelled responses of the near-lock detector

Exact performance has been accurately modelled, based on the use of a simple first order low-pass filter, and the story begins in the modelling of the filter response to a variety of sweep rates. When the loop is swept, the phase detector output consists of a beat note going through DC when the reference and VCO frequencies are instantaneously equal, at time $t = 0$, say. The filter response depends on two factors: sweep rate and relative phase – several examples of which are shown in figure 6.6. Clearly, a higher sweep rate results in a smaller filter response (for a given cut-off frequency) but there is a considerable – and perhaps surprising – dependence of the peak response on the relative phase, ϕ_o, of the two signals. This relative phase is an entirely random quantity since the VCO and input signals are completely uncorrelated prior to lock and thus there is a statistical variation in the peak responses of the low-pass filters from sweep to sweep. To model this, a large number of responses of the type shown in figure 6.6 were computed and the peak filter response was recorded as a function of both the sweep rate and relative phase, the results of which are plotted in figure 6.7. As expected, a higher sweep rate results in a reduced peak filter response along with a very significant dependence on the random variable ϕ_o. In fact, the peak response varies with ϕ_o by a factor of between two and three. From figure 6.7 it is very easy to derive a probability density function for a given threshold setting.

Results applying to a nominal threshold setting of 50% are shown in figure 6.8. If only one detector is used then the detection probability is equal to the proportion of the given plot in figure 6.7 which lies above the threshold. For example, from figure 6.7, a relatively low sweep rate of $K/\omega_o^2 = 0.4\pi$ gives a peak response above the 50% threshold for roughly 90% of all possible values of ϕ_o and so the detection probability is approximately 90%. The advantage of using two separate detectors operating on both the quadrature-driven and in-phase-driven phase detectors is that detection is achieved if either the peak response as a function of ϕ_o or the peak response as a function of ($\phi_o - 90°$) exceeds 50%; the minor disadvantage being that there is a slightly greater chance of false detection due to noise. Thus circuit behaviour with two detectors is far superior, with 100% detection probability for sweep rates of up to $K/\omega_o^2 = 2.2\pi$. Figure 6.8 also shows some experimental points which were obtained with the prototype system indicating good general agreement with the modelling. The final parameter of interest is the response time of the near-lock detector. Intuitively, an earlier response would be expected from a lower sweep rate, along with some variation with ϕ_o – and this is indeed the case. Figure 6.9 shows that the response time varies over the range $-0.9/\omega_o$ to $+0.75/\omega_o$ s for normalised sweep rates of between 0.4π and 2π.

From the results of figures 6.6 to 6.9 it is easy to arrive at a viable design. Firstly, from figure 6.8, $2\pi\omega_o^2$ rad/s^2 is just about the highest sweep rate giving 100% probability of detection on a single sweep and

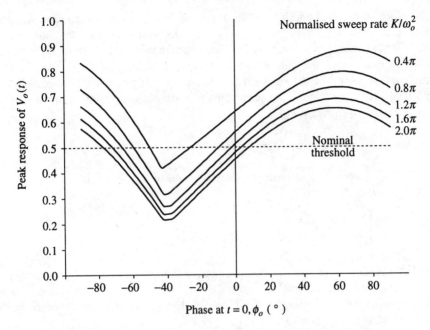

Figure 6.7 Peak response of the near-lock detector with ϕ_o and sweep rate

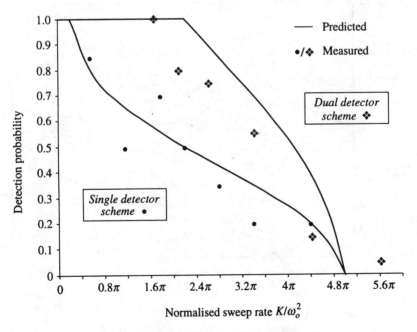

Figure 6.8 Detection probability variation with sweep rate

therefore seems a good choice. The response time for this sweep rate, from figure 6.9, is up to $0.75/\omega_o$ s which corresponds to an overshoot of $(0.75/\omega_o) \times (2\pi\omega_o^2) = 1.5\pi\omega_o$ rad/s. This overshoot must be less than the loop capture range for the loop to acquire lock. Taking a safe estimate for the capture range of a loop with an analogue multiplier phase detector of ± 5 times the loop natural frequency gives a simple design equation for the filter cut-off frequency ω_o,

$$1.5\pi\omega_o = 5\omega_n \quad \text{i.e.} \quad \omega_o = 1.06\omega_n$$

and so the sweep rate in terms of the loop natural frequency is

$$\left[\frac{d\omega}{dt}\right]_{\text{max}} = 7.1\omega_n^2 \quad \text{rad/s}^2 \tag{6.8}$$

This represents a 14-fold improvement over the sweep rate usable with the more conventional scheme of figure 6.3 and, if careful attention is paid to phase detector DC offsets, significantly higher sweep rates could be supported with this method.

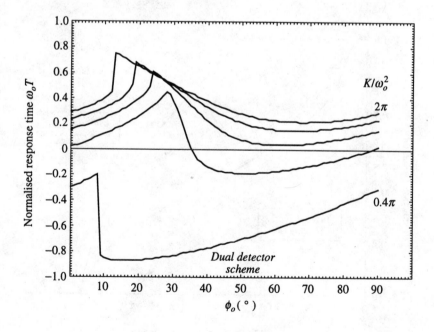

Figure 6.9 Predicted response time of the near-lock detector

Although this analysis has been based on a PLL incorporating a multiplier phase detector with a sinusoidal characteristic, the results may be applied with good accuracy to a PLL using an exclusive-OR type of phase detector with a modified threshold of 41% rather than 50%. The change in threshold arises because an exclusive-OR phase detector produces a triangular beat note (of peak amplitude A, say) of which the low-pass filters pass substantially the fundamental component having an amplitude of $8A/\pi^2$ – thus requiring a reduction in threshold to $8/\pi^2$ of its former value for similar performance. Of course, this argument is somewhat approximate, although it has been successfully confirmed in practice.

7 Higher Order Loop Filters

7.1 Supplementary filtering

The second order type II loop filter considered quite extensively until now is quite commonly used in PLL circuits because of its simplicity, inherent stability and ease of analysis. However, it does not offer very good rejection of high frequency signals from the phase detector, the gain of the filter itself – from figure 3.1(b) – being a constant value of R_2/R_1 at high frequencies. Thus the trade-off between loop bandwidth and reference sideband levels is not particularly good. Improvements can be made either by using a higher order filter design or by including additional filtering in an existing second order loop design. In the latter case, the filtering should be designed to have little effect on the loop characteristics within or close to the loop natural frequency, so that stability is not impaired, yet it should introduce some useful suppression at higher frequencies so that reference sideband levels may be reduced.

Figure 7.1 shows some examples of how additional filtering may be included. The input resistor may be split into two of equal value, $R_1/2$, and a capacitor, C_4, connected to ground to introduce a further pole into the transfer characteristic. This practice is recommended for certain types of phase detector where the output pulses are too narrow for the op-amp to detect. The response of the additional filtering is simply

$$H(s) = \frac{\omega_o}{s + \omega_o}, \text{ where } \omega_o = \frac{4}{R_1 C_4} \tag{7.1}$$

Of course, a simple RC filter could be introduced at some other point in the circuit between the phase detector and VCO with similar effect. This filtering introduces a phase shift which degrades the loop phase margin and may cause instability if its cut-off is too low. It is important, therefore, that the phase shift introduced by the filter is small compared to the loop phase margin at the frequency, ω', at which the loop gain falls to unity. From chapter 3, the loop gain is unity when $(\omega'/\omega_n)^2 = 2\zeta^2 + \sqrt{(4\zeta^4 + 1)}$ where ω' along with the phase margin are tabulated for several values of ζ as shown in table 7.1.

Table 7.1 Phase margin and ω' versus damping factor for a
second order type II PLL

ζ	0.5	0.707	1
ω'/ω_n	1.28	1.55	2.06
Phase margin	51.8°	65.5°	76.3°

A sensible criterion when including additional filtering in a second
order PLL design is that the phase shift introduced by the filtering,
$\phi = \tan^{-1}(\omega/\omega_o)$ in this case, should be no more than 10° at frequency
$\omega = \omega'$; in other words the phase margin is degraded by no more than 10°.
Taking the case of a damping factor of 0.707, this means that the filter
cut-off frequency should be at least $\omega_o = \omega'/\tan 10° = (\omega'/\omega_n)(\omega_n/\tan 10°) =$
$1.55\omega_n/\tan 10° = 8.8\omega_n$. Such high filter cut-offs are of little use in
rejecting noise close to the loop natural frequency but are helpful in
rejecting noise of significantly higher frequency. It is also worth noting
that if the filter cut-off frequency is $\omega_o = 0.66\omega_n$ then the phase margin is
zero and the loop is on the verge of instability.

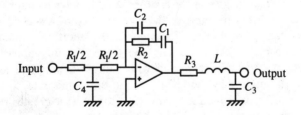

Figure 7.1 General example of a higher order loop filter

Improved rejection can be obtained by using a two-pole LCR filter, as
shown in figure 7.1. The transfer function in this case can easily be shown
to be

$$H(s) = \frac{1/LC_3}{s^2 + (R_3/L)s + 1/LC_3} \tag{7.2}$$

and this may be written in the familiar form,

$$H(s) = \frac{\omega_o^2}{s^2 + 2\zeta\omega_o s + \omega_o^2}, \text{ where } \omega_o = \frac{1}{\sqrt{LC_3}} \text{ and } \zeta = \frac{R_3}{2}\sqrt{\frac{C_3}{L}}$$

In this expression, ω_o represents the filter cut-off frequency and ζ a shaping factor (not to be confused with the loop damping factor). For a shaping factor of 0.707 and, again, assuming that the phase margin may be eroded by no more than 10°, the filter cut-off frequency must now be at least $\omega_o = 12.6\omega_n$. In comparison with RC filtering, this filter has a sharper response but a slightly higher cut-off frequency – giving greater suppression of noise beyond approximately 20 times the loop natural frequency. It is easy to evaluate the amount of reference sideband suppression obtainable with this method. Taking, for example, the case of a loop with a natural frequency equal to 1% of the reference frequency, the suppression offered by RC filtering is around 100/8.8 or 21.1 dB whereas the suppression offered by LCR filtering is around $(100/12.6)^2$ or 36.0 dB.

7.2   The third order type II loop filter

Rather than cautiously applying additional filtering to a basic lower order design it is preferable to redesign the loop filter from scratch to give a higher order loop. A further pole may be introduced into a second order type II design by bypassing resistor R_2 with a capacitor C_2, as shown in figures 7.1 and 7.2, to give a third order type II loop characteristic. This technique may be used to improve the performance of a second order loop filter or preferably, by means of designing the additional component in, may be used to create a third order loop filter. The advantage of a third order loop over a second order loop with additional supplementary filtering is that, although they both roll-off at the same rate, the response begins to roll-off sooner with a third order loop so that greater rejection of loop noise components is possible. This means that, for a given loop natural frequency, the reference sidebands are lower.

The transfer function of this loop filter design is

$$G(s) = \frac{1/sC_1 + R_2/(1 + sC_2R_2)}{R_1} \equiv \frac{1}{sC_1R_1}\left[\frac{1 + s(C_1 + C_2)R_2}{1 + sC_2R_2}\right] \tag{7.3}$$

which, as a shorthand, may be written as,

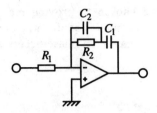

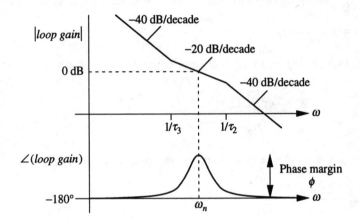

Figure 7.2 Third order type II loop filter and associated Bode plots

$$G(s) = \frac{1}{s\tau_1}\left[\frac{1+s\tau_3}{1+s\tau_2}\right]$$ (7.4)

where $\tau_1 = R_1 C_1$, $\tau_2 = R_2 C_2$ and $\tau_3 = R_2(C_1 + C_2)$. The design strategy for this filter is quite different to that of the familiar second order filter: τ_1 is chosen so that the loop gain, $K_p K_v G(s)/N$, falls to unity at the loop natural frequency, ω_n, and τ_2 and τ_3 are chosen to provide the required phase margin, ϕ. The piecewise linear bode plot of the loop gain (which therefore includes the $1/s$ VCO term) is as shown in figure 7.2. The loop gain has a low-pass characteristic with a cut-off rate of 40 dB/decade at frequencies below $1/\tau_3$ and above $1/\tau_2$ and a roll-off rate of 20 dB/decade in between. The phase shift is 180° at frequencies well below $1/\tau_3$ and well above $1/\tau_2$ and is somewhat less than 180° in between. The phase of the loop gain $K_p K_v G(s)/Ns$, from equation (7.4), is given by

$$\angle loop\ gain = -\pi + \tan^{-1}\left[\frac{\omega(\tau_3 - \tau_2)}{1+\omega^2 \tau_2 \tau_3}\right]$$ (7.5)

where τ_2 and τ_3 must be chosen to provide maximum phase margin at $\omega = \omega_n$. By equating the differential of the argument of the $\tan^{-1}$ function to zero, it is found that the maximum phase occurs at a frequency given by

$$\omega_n^2 = 1/\tau_2\tau_3 \tag{7.6}$$

indicating that the loop natural frequency is the geometric mean of $1/\tau_2$ and $1/\tau_3$ – an intuitively obvious result. The phase margin is thus given by

$$\phi = \angle loop\ gain(\omega_n) + \pi$$

which, substituting equation (7.6) into equation (7.5), results in

$$\phi = \tan^{-1}\left[\frac{\tau_3 - \tau_2}{2\sqrt{\tau_2\tau_3}}\right] \tag{7.7}$$

Eliminating τ_2 in the last two expressions gives a quadratic in τ_3,

$$\tau_3^2 - \frac{2\tan\phi}{\omega_n}\tau_3 - \frac{1}{\omega_n^2} = 0$$

with the solution

$$\tau_3 = \frac{\tan\phi + \sec\phi}{\omega_n} \tag{7.8}$$

Since $\tau_2 = 1/(\omega_n^2\tau_3)$ then

$$\tau_2 = \frac{1}{\omega_n(\tan\phi + \sec\phi)} \tag{7.9}$$

So, for a desired phase margin, ϕ, τ_2 and τ_3 can be found. It now remains to find an expression for τ_1. This is achieved by considering the magnitude of the loop gain at the loop natural frequency ω_n,

$$|loop\ gain| = \frac{K_pK_v}{N\omega_n^2\tau_1}\sqrt{\frac{1 + \omega_n^2\tau_3^2}{1 + \omega_n^2\tau_2^2}} \equiv \frac{K_pK_v}{N\omega_n^2\tau_1}(\tan\phi + \sec\phi)$$

and this must be unity so that maximum phase margin is obtained, giving

$$\tau_1 = \frac{K_pK_v}{N\omega_n^2}(\tan\phi + \sec\phi) \tag{7.10}$$

The filter design proceeds directly from these results:

$$R_1 C_1 = \frac{K_p K_v}{N \omega_n^2}(\tan\phi + \sec\phi) \tag{7.11}$$

$$R_2 C_2 = \frac{1}{\omega_n (\tan\phi + \sec\phi)} \tag{7.12}$$

and

$$R_2 C_1 = \tau_3 - \tau_2 = \frac{1}{\omega_n}\left[\tan\phi + \sec\phi - \frac{1}{\tan\phi + \sec\phi}\right]$$

which, by applying standard trigonometric identities, reduces to

$$R_2 C_1 = \frac{2\tan\phi}{\omega_n} \tag{7.13}$$

So, although the analysis is more complex, the design process is almost as simple for a third order filter as it is for a second order filter. However, because the peak in the filter phase response must occur close to the frequency where the loop gain is unity, the design is more sensitive to errors in the estimation of loop parameters or, in synthesiser applications, in variations in their values with division ratio, and this point should be borne in mind when considering a practical circuit design.

7.3   Performance of the third order type II loop

7.3.1 Modulation

The phase modulation characteristic of the loop may be found by substituting the new filter response, $G(s)$, into the general expression for the closed loop transfer function, from equation (2.11),

$$\frac{\phi_o(s)}{\phi_i(s)} = \frac{\dfrac{K_p K_v}{s^2 \tau_1}\left[\dfrac{1 + s\tau_3}{1 + s\tau_2}\right]}{1 + \dfrac{K_p K_v}{N s^2 \tau_1}\left[\dfrac{1 + s\tau_3}{1 + s\tau_2}\right]} \equiv \frac{N\left[1 + \dfrac{(\tan\phi + \sec\phi)s}{\omega_n}\right]}{\dfrac{s^2}{\omega_n^2}(\tan\phi + \sec\phi) + \dfrac{s^3}{\omega_n^3} + 1 + \dfrac{(\tan\phi + \sec\phi)s}{\omega_n}}$$

which simplifies to,

$$\frac{\phi_o(s)}{N\phi_i(s)} \equiv \frac{K_p\phi_o(s)}{NV_{pm}(s)} = \frac{\omega_n^2(\tan\phi + \sec\phi)s + \omega_n^3}{s^3 + \omega_n(\tan\phi + \sec\phi)s^2 + \omega_n^2(\tan\phi + \sec\phi)s + \omega_n^3}$$

$$(7.14)$$

This demonstrates a third order low pass response (since the denominator is a third order polynomial), with a high frequency roll-off of 40 dB/decade. The magnitude of this characteristic is plotted in figure 7.3. The response has a peak at around $0.8\omega_n$ which is more pronounced at lower phase margins. This peak has a magnitude of 5.7 dB, 3.1 dB and 1.7 dB for phase margins of 30°, 45° and 60°, respectively, the latter two values being most useful in practical designs. Because of the 40 dB/decade high frequency roll-off rate, the filter offers superior suppression of phase noise beyond the loop natural frequency compared to a second order filter and this means that reference sideband levels are substantially lower. The improved suppression allows a higher loop natural frequency to be used for a given phase detector operating frequency and generally enables a better trade-off between noise performance and tracking agility to be obtained.

The frequency modulation characteristic can be found in a similar manner to that used for the second order loop resulting in

$$\frac{\omega_o(s)}{K_v V_{fm}(s)} = \frac{1}{1 + \dfrac{K_p K_v G(s)}{Ns}}$$

$$= \frac{s^3 + \omega_n(\tan\phi + \sec\phi)s^2}{s^3 + \omega_n(\tan\phi + \sec\phi)s^2 + \omega_n^2(\tan\phi + \sec\phi)s + \omega_n^3} \quad (7.15)$$

This is a third order high-pass response with a roll-off of 40 dB/decade at low frequencies, as plotted in figure 7.4. It is slightly unfortunate that the response contains a peak for all values of phase margin – unlike the second order filter which may be designed for a Butterworth characteristic. It is also interesting to note that the PM and FM characteristics appear to be mirror images of each other about $\omega = \omega_n$. In fact it is easy to show that this is the case by applying the standard low-pass to high-pass transformation, in which s is replaced by ω_n^2/s, to equation (7.15),

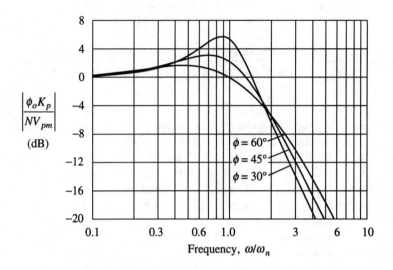

Figure 7.3 Phase modulation characteristic of a third order type II PLL

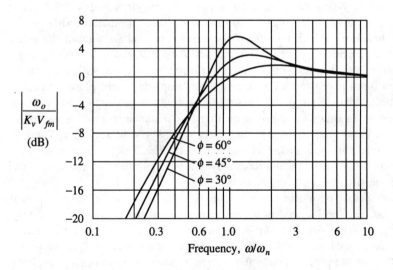

Figure 7.4 Frequency modulation characteristic of a third order type II PLL

$$\frac{\omega_o(\omega_n^2/s)}{K_v V_{fm}(\omega_n^2/s)} = \frac{\omega_n^6/s^3 + (\tan\phi + \sec\phi)\omega_n^5/s^2}{\omega_n^6/s^3 + (\tan\phi + \sec\phi)\omega_n^5/s^2 + (\tan\phi + \sec\phi)\omega_n^4/s + \omega_n^3}$$

$$\equiv \frac{\omega_n^2(\tan\phi + \sec\phi)s + \omega_n^3}{s^3 + \omega_n(\tan\phi + \sec\phi)s^2 + \omega_n^2(\tan\phi + \sec\phi)s + \omega_n^3} = \frac{K_p\phi_o(s)}{NV_{pm}(s)}$$

$$(7.16)$$

and this proves that the PM and FM characteristics are symmetrical about $\omega = \omega_n$ for a third order loop filter.

Two-point modulation may be used with the third order filter in exactly the same way as for the second order filter where, by adding the two characteristics of equations (7.14) and (7.15) together, it is quite clear that the combination of PM and FM modulation gives rise to a flat response.

7.3.2 Transient response

The main benefit of the third order type II filter is that it offers a considerable reduction in reference sidebands, allowing a better compromise between tuning speed and reference sidebands to be reached. This is of particular importance in frequency synthesiser applications, for which the author would very much recommend the use of this particular loop filter design. When comparing the dynamic behaviours of second order type II and third order type II loops there is no appreciable difference. This is because both filters are of the same type and so contain the same number of supposedly perfect integrators. Any PLL using a type II filter, irrespective of its order, responds to an input phase step with zero phase error, an input phase ramp (i.e. a frequency step) with zero phase error and a parabolic change in input phase (i.e. a frequency ramp) with a finite phase error. In order to improve on this, for instance to track a parabolic change in input frequency, a higher filter *type* would be required. There is considerable variation in the classification of PLL filter characteristics in PLL texts which may be overcome by thinking in terms of the order and type of the filter.

The response of a third order type II loop to a step change in input phase, $\Delta\phi$, may be derived in a similar manner to that shown for a second order type II loop in chapter 3. From the closed-loop transfer function arrived at in equation (7.14) and considering the case of a phase margin of 53.2° (for mathematical convenience and because it represents a realistic choice), the output phase transient in Laplace notation is given by the following,

$$\frac{\phi_o(s)}{N} = \Delta\phi\frac{3\omega_n^2 s + \omega_n^3}{s(s+\omega_n)^3}$$

and, resolving into partial fractions, this is equivalent to

$$\frac{\phi_o(s)}{N} = \Delta\phi\left[\frac{1}{s} - \frac{1}{s+\omega_n} - \frac{\omega_n}{(s+\omega_n)^2} + \frac{2\omega_n^2}{(s+\omega_n)^3}\right] \tag{7.17}$$

Taking the inverse Laplace transform, the loop response in the time domain is therefore

$$\frac{\phi_o(t)}{N} = \Delta\phi\left[1 + \left[(\omega_n t)^2 - \omega_n t - 1\right]e^{-\omega_n t}\right] \tag{7.18}$$

and so the phase error present at the phase detector is given by

$$\phi_e(t) = \Delta\phi - \frac{\phi_o(t)}{N} = \Delta\phi\left[1 + \omega_n t - (\omega_n t)^2\right]e^{-\omega_n t} \tag{7.19}$$

This function is plotted in figure 7.5 and is of very similar form to that of the second order type II filter. However, it is noticeable that the settling time is approximately 50% greater so that, for comparable transient performance, the loop natural frequency needs to be slightly higher. It is important to remember that the phase and frequency errors referred to in the analysis are those experienced at the phase detector inputs and therefore must be scaled by a factor of N in order to relate to the VCO output.

The loop response to an input frequency step, $\Delta\omega$, is easily found by integrating equation (7.19) and multiplying by $\Delta\omega/\Delta\phi$. Using integration by parts the result is

$$\phi_e(t) = \frac{\Delta\omega}{\omega_n}\left[\omega_n t + (\omega_n t)^2\right]e^{-\omega_n t} \tag{7.20}$$

As before, for the second order type II loop, the phase error is initially zero, then increases to a peak and eventually declines to zero, as plotted in figure 7.6(a). The peak phase error may be found from the turning points of equation (7.20) which may be achieved by setting equation (7.19) to zero. This leads to the conclusion that the maximum phase error occurs at $\omega_n t = (1 + \sqrt{5})/2$ and, substituting in equation (7.20), has a value of

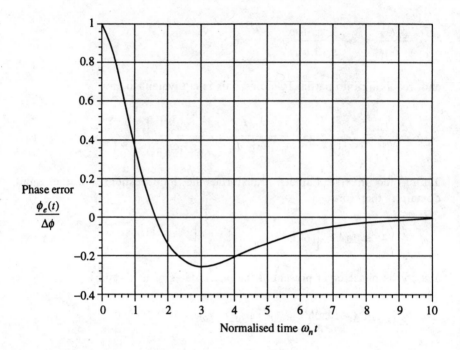

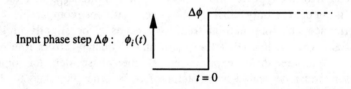

Figure 7.5 Response of a third order type II PLL ($\phi = 53.2°$) to a phase step

$$\phi_e(\text{peak}) = 0.84\frac{\Delta\omega}{\omega_n} \qquad (7.21)$$

This compares with a peak phase error of $\Delta\omega/e\omega_n$ for the second order loop, showing that a greater phase error is to be expected for a given frequency step. Indeed, comparison of the phase error plots for second and third order loops reveals that the third order loop has slightly inferior performance. Regardless of the filter design, the loop behaves in a linear fashion only when the peak phase error remains within the phase detector range and so there is considerable benefit in using extended range phase detectors where possible (see chapter 9).

The final item of interest is the frequency error transient which is simply the differential of equation (7.20),

$$\omega_e(t) = \Delta\omega\left[1 + \omega_n t - (\omega_n t)^2\right]e^{-\omega_n t} \qquad (7.22)$$

This function is plotted in figure 7.6(b) and, of course, is of similar form to the phase step transient of figure 7.5. The preceding analysis may be used to determine whether loop operation is likely to be linear and, if so, the exact form of the resulting transient. In frequency synthesiser applications, an analysis of this form may be of immense use in predicting the tuning characteristic when changing between channels. Example 9 forms a useful illustration of the limits on the linear range of a PLL with a 3rd order type II filter and is also an interesting comparison of the behaviour of second order and third order loops in this respect.

7.3.3 *Reference sidebands*

Reference sidebands arise due to a small proportion of the inevitable AC component of the phase detector output passing through the loop filter and phase modulating the VCO. With sequential phase detectors, the fundamental component of the modulation – and hence the sideband spacing – is at the reference frequency, whereas with multiplier phase detectors it is at twice the reference frequency. The addition of any spurious component on the output of a frequency synthesiser is of course undesirable, but reference sidebands are a particular nuisance because they fall on adjacent channels which, in communications applications, may give rise to adjacent channel interference.

PLLs incorporating phase/frequency detectors have relatively low reference sidebands because the phase detector output has no AC component, in principle, when the loop is locked. However, in practice, there are a number of mechanisms producing reference sidebands – such as switching transients from the phase detector coupling onto the supply rails and appearing elsewhere in the loop. Probably the most serious cause of reference sidebands is the presence of DC offsets in the phase detector or loop filter. Any DC offset causes the loop to adjust in order to restore equilibrium, requiring a small phase difference between the phase detector inputs which produces short pulses on one of the phase/frequency detector outputs, as shown in figure 7.7. Of course, the mean value of the pulsed output is equal and opposite to the DC offset and so the duty cycle (i.e. ratio of pulse width, Δt, to reference period, T_{ref}) of such pulses may easily be related to the DC offset, ΔV,

$$\frac{\Delta t}{T_{ref}} = \frac{\Delta V}{2\pi K_p} \qquad (7.23)$$

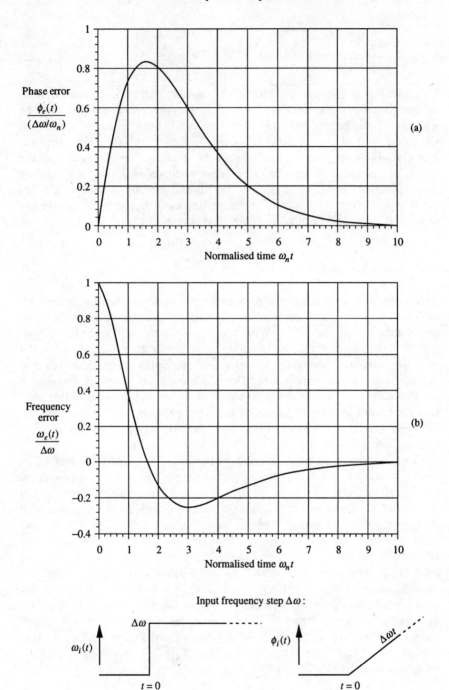

Figure 7.6 Response of a third order type II PLL ($\phi = 53.2°$) to a frequency step

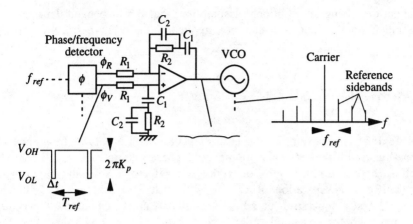

Figure 7.7 Formation of reference sidebands

Since the peak phase detector output voltage is $2\pi K_p$ then the percentage duty cycle is simply equivalent to the percentage DC offset – around 0.1% in a good design.

The reference sideband distribution may be found by Fourier analysis of the phase detector pulses, of amplitude $2\pi K_p$ and duration Δt, where the amplitude of the nth harmonic, of frequency nf_{ref}, is

$$n\text{th } harmonic\ amplitude\ =\ 4K_p\,\frac{\sin(n\pi\Delta t/T_{ref})}{n}$$

Because phase detector ripple appears at the same point in the loop at which phase modulation is inserted, then it may be treated as a source of phase modulation by a number of sinusoidal components with amplitudes given by the previous equation. Thus the response of a third order type II loop may be obtained with the aid of equation (7.14),

$$Peak\ phase\ deviation\ =\ 4K_p\,\frac{\sin(n\pi\Delta t/T_{ref})}{n}\ \times$$

$$\frac{N}{K_p}\left|\frac{\omega_n^2(\tan\phi+\sec\phi)s+\omega_n^3}{s^3+\omega_n(\tan\phi+\sec\phi)s^2+\omega_n^2(\tan\phi+\sec\phi)s+\omega_n^3}\right|_{s=jn\omega_{ref}}$$

Assuming that the total phase deviation due to this effect is small, then a narrow-band PM approximation may be employed (appendix B) in which sideband levels relative to the carrier are equal to half the peak phase

deviation. Using an additional assumption, that the loop natural frequency
is much smaller than the reference frequency, the relative sideband levels
are:

$$\textit{Reference sideband levels} = \frac{2N}{n^3}\sin(n\pi\Delta t/T_{ref})\left[\frac{\omega_n}{\omega_{ref}}\right]^2 (\tan\phi + \sec\phi) \quad (7.24)$$

Taking, as an example, the case of a PLL with a natural frequency of
one hundredth the reference frequency, a phase margin of 60°, a 0.1% DC
offset and a divider ratio of 1000, the reference sideband levels are
−53 dBc at a separation of $\pm f_{ref}$, −65 dBc at a separation of $\pm 2f_{ref}$ and
−72 dBc at a separation of $\pm 3f_{ref}$. A similar result to equation (7.24) may
be obtained for a second order type II loop, where it is found that the
reference sideband levels are around 30 dB higher − providing convincing
evidence of the merits of a third order design.

The reference sideband levels of this example are a little on the high
side, though it is easy to improve matters by including some additional
filtering, as described in section 7.1, as long as this filtering has little
impact on loop stability and general dynamic performance. In particular, a
carefully designed LCR filter placed between the loop filter and VCO is
highly effective at reducing reference sidebands. Also, since the filter is
placed directly before the VCO tuning voltage terminal, it suppresses
many other sources of noise present in the circuit. From section 7.1, an
LCR filter with a cut-off frequency of 12.6 times the loop natural
frequency is a relatively safe choice (degrading the phase margin of a
second order type II loop by only 10°) and attenuates signals at the
reference frequency (around 100 times the loop natural frequency) by
approximately a factor of $(100/12.6)^2$ or 36 dB. So, a considerable reduction in reference sidebands is possible by the judicious use of such
additional filtering.

8 Digital Loop Techniques and Design Methods

8.1 Fractional-N synthesis

Classical frequency synthesisers are based on PLLs with programmable dividers in the feedback path and achieve a range of output frequencies in multiples of the reference frequency. Naturally enough, in order to achieve high resolution, a low reference frequency is required. Since a low reference frequency implies a low loop natural frequency then there is clearly a trade-off in such synthesisers between resolution on the one hand and transient response and tuning speed on the other. If a synthesiser is to be designed for an application where high resolution and fast tuning are both important, then it is tempting to increase the natural frequency beyond the reliable $f_{ref}/100$ limit – perhaps to as much as $f_{ref}/10$. This action, however, serves to greatly increase the reference sidebands and provide an uncomfortable compromise in overall loop performance.

Fractional-N synthesisers are a very different approach and operate by achieving the effect of a loop with a *fractional* division ratio in the feedback chain. This means that a high resolution is possible without the need for a very low reference frequency and the associated limitation on loop agility. The concept of the technique is very simple although, as always, the details are slightly more complicated. Referring to figure 8.1, fractional-N synthesisers are based on a conventional synthesiser design, but with additional control circuitry to rapidly alternate the division ratio between N and $(N + 1)$ under the control of two digital inputs, N and p which set the integer and fractional components of the division ratio, respectively. Depending on the proportion of divisions by either ratio, the loop is able to multiply the reference frequency by some amount between N and $(N + 1)$. For instance, if the division ratio alternates equally between N and $(N + 1)$ then an average division ratio of $(N + 1/2)$ is achieved and so the output frequency is $(N + 1/2)$ times the reference frequency. Standard digital techniques may be used to control the programmable divider to provide a range of average division ratios. In the circuit of figure 8.1, an n-bit accumulator is clocked by the output of the usual loop divider and each time increases its stored value by an amount p. Denoting the accumulator size by q, where $q = 2^n$, then the accumulator overflows, on average, during p/q of the divider output cycles. An overflow signal is derived from the accumulator which increments the programmable divider to $(N + 1)$ each time an overflow occurs, producing an average division ratio of $(N + p/q)$ and thus a mean synthesiser output frequency of

$$f_c = \left[N + \frac{p}{q} \right] f_{ref} \qquad (8.1)$$

The accumulator size is chosen to suit the desired resolution. For instance, for a resolution of 1/16th of the reference frequency, a 4-bit accumulator is needed.

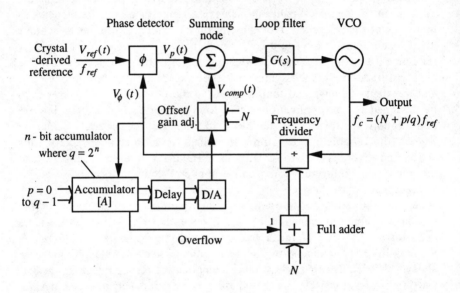

Figure 8.1 A fractional-N synthesiser

A major drawback of the system as described so far is that the alternating action of the frequency divider, in effect, phase modulates the feedback signal to the phase detector. This phase modulation, which may have a fundamental frequency as low as f_{ref}/q (when $p = 1$ or $q - 1$), causes AC signal components on the phase detector output which give rise to sidebands (fractional-N sidebands) on the VCO output signal. As we shall see, these sidebands may be very large and so the problem is addressed by a cunning open-loop compensation method shown in figure 8.1 and described as follows.

Figure 8.2 illustrates the operation of a fractional-N synthesiser including the use of compensation to suppress fractional-N sidebands. The figure

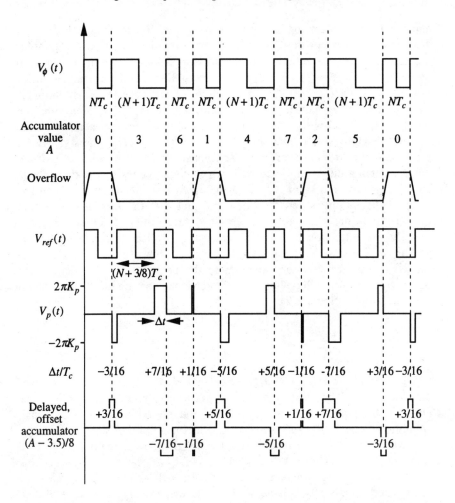

Figure 8.2 Operation of a fractional-N synthesiser with $q = 8$, $p = 3$ and $N = 1$

applies to the case of a system with $q = 8$ and $p = 3$ which provides a fractional division ratio component of 3/8. For clarity, the integer component of the division ratio, N, has been taken as 1 though, in practice, a very much higher value is likely. To begin with, on the first cycle of the divider output, $V_\phi(t)$, the accumulator value, A, is 0, having just over-flowed, and the division ratio is N. On the second cycle of the divider output, the accumulator value increases by an amount p to 3 and the previous overflow condition results in a division ratio of $(N + 1)$. On the third cycle, the accumulator value increases to 6 and the division ratio is N. On the fourth cycle, the accumulator value overflows to 1 and the division ratio is N. On the fifth cycle, the accumulator value increases to 4 and the previous overflow condition results in a division ratio of $(N + 1)$. This pattern continues until, after 8 cycles of divider output, the accumulator value returns to zero and the process repeats. It is evident that the division ratio is $(N + 1)$ for 3 cycles in every 8 and so the average division ratio is indeed $(N + 3/8)$. The phase detector response to the situation is easy to derive by plotting the reference frequency, V_{ref}, alongside the divider output. This is shown in figure 8.2 where the period of the reference signal, from equation (8.1), is $T_{ref} = (N + p/q)T_c$. The signal $V_p(t)$ represents the response of a phase/frequency detector with a tri-state output, pulsing either high or low between rising edges of the two inputs, depending on which input is leading the other. It is clear that the mean DC level of the output is zero – which indicates that the loop is in a stable locked condition – but there is also a considerable AC component due entirely to the fractional-N process, consisting of discrete pulses of various widths Δt. These pulse widths vary over the range $\pm(q - 1)T_c/2q$ and are related to the accumulator value, A, in the period *preceding* that of the pulse by the following function (which may be deduced by careful consideration of figure 8.2):

$$\frac{\Delta t}{T_c} = \frac{(q-1)/2 - A}{q} \tag{8.2}$$

For example, $q = 8$, $A = 0$ gives $\Delta t/T_c = 7/16$ and $q = 8$, $A = 3$ gives $\Delta t/T_c = 1/16$. The maximum pulse width of these undesirable phase detector components is just below half the period of the VCO signal, which may actually be very small (e.g. 1 ns for 500 MHz VCO frequency); however, this is still sufficient to seriously degrade spectral purity if no form of compensation is included – as shown later in this section. Fortunately, it is relatively easy to provide such compensation since the pulse width is directly related to A. As is evident in the bottom plot of figure 8.2, the delayed accumulator value, offset by $(q - 1)/2$ and scaled by $1/q$, exactly mirrors the pulse width variation of the phase detector output and so an offset, scaled version of the accumulator value may be used as compensation, either to adjust the phase of the reference signal or to apply a correcting voltage to the phase detector output. The latter method was originally developed and patented by the Racal company and is shown in figure 8.1. Since the period of the signals at the phase detector inputs is very nearly equal to NT_c (assuming N is large), then the mean phase detector output voltage arising from an individual pulse is $2\pi K_p \Delta t/(NT_c)$ and so the required compensating voltage is

$$V_{comp} = -\frac{2\pi K_p}{N}\frac{\Delta t}{T_c} = \frac{\pi K_p(2A - q + 1)}{Nq} \tag{8.3}$$

The $(-q + 1)$ term in this expression represents an offset which may easily be implemented by simply AC coupling the compensation onto the phase detector output. It is important to note that the compensation must be scaled by a factor of $1/N$ in order to be accurate over a range of output frequencies. This scaling may best be achieved digitally, before the D/A convertor of figure 8.1. It must also be realised that this is an entirely open-loop technique and so relies heavily on stable, accurate and linear circuit techniques in the analogue regions of the circuit (basically, around the phase detector).

We shall now consider the amplitude of the worst-case sidebands due to the *uncompensated* fractional-N process. Figure 8.3 shows behaviour with a division ratio of $N + 1/8$. The frequency divider now divides by $(N +1)$

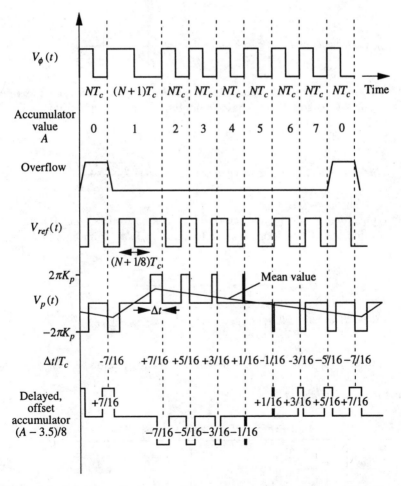

Figure 8.3 Operation of a fractional-N synthesiser with $q = 8$, $p = 1$ and $N = 1$

for one cycle in every 8, which leads to a sawtooth-like phase modulation in the feedback path to the phase detector which repeats every q cycles of the reference frequency. The situation is therefore exactly equivalent to a PLL locked to a reference signal containing sawtooth phase modulation. The frequency of this phase modulation may, unfortunately, be much lower than the loop reference allowing easy propagation through the loop filter and subsequent modulation of the VCO. For a system with moderately high resolution, $q \geq 16$ say, the apparent phase modulation caused by the process approximates closely to a sawtooth function of frequency f_{ref}/q and of peak value just below π/N radians. Fourier analysing this sawtooth waveform, the amplitude of the nth harmonic is $2/(nN)$ radians and so the amplitude of each harmonic of the ensuing *VCO* phase modulation may be easily found:

$$Peak\ phase\ deviation\ of\ nth\ harmonic\ =\ \frac{2}{n}\left|\frac{\phi_o(nf_{ref}/q)}{N\phi_i(nf_{ref}/q)}\right|$$

and assuming the overall phase deviation is quite small so that the linear behaviour of narrow-band phase modulation applies, then the sideband levels are given by

$$Sideband\ amplitudes\ relative\ to\ carrier\ =\ \frac{1}{n}\left|\frac{\phi_o(nf_{ref}/q)}{N\phi_i(nf_{ref}/q)}\right| \tag{8.4}$$

Applying this result to the case of a second order type II loop, by substituting the closed loop transfer function, $\phi_o/N\phi_i$, from equation (4.2a), gives

$$\begin{matrix} Fractional\text{-}N \\ sideband\ levels \end{matrix} = \frac{1}{n}\sqrt{\frac{1+4\zeta^2(nf_{ref}/qf_n)^2}{\left[1-(nf_{ref}/qf_n)^2\right]^2+4\zeta^2(nf_{ref}/qf_n)^2}} \tag{8.5}$$

and similarly for a third order type II loop, from equation (7.14),

Fractional - N sideband levels =

$$\frac{1}{n}\sqrt{\frac{1+(nf_{ref}/qf_n)^2(\tan\phi+\sec\phi)^2}{\left[\left[\frac{nf_{ref}}{qf_n}\right]^3-\left[\frac{nf_{ref}}{qf_n}\right](\tan\phi+\sec\phi)\right]^2+\left[1-\left[\frac{nf_{ref}}{qf_n}\right]^2(\tan\phi+\sec\phi)\right]^2}} \tag{8.6}$$

These expressions have been derived without the simplifying assumption that the sawtooth modulation frequency is much greater than the loop natural frequency since this may not be a valid assumption in this case. For a realistic idea of the magnitude of these sidebands, we shall consider the case of a system with a loop natural frequency of $f_n = f_{ref}/100$ (sufficient for *reference* sidebands of around −60 dBc in a practical design), a resolution of 1/16th of the reference frequency (i.e. $q = 16$), a damping factor of 0.707 for a second order loop and a phase margin of 60° for a third order loop. The first pair of sidebands, at a frequency separation of $\pm f_{ref}/16$ from the carrier, from equations (8.5) and (8.6), have a level of 0.23, or −13 dBc, for the second order loop and 0.09, or −21 dBc, for the third order loop. These values are very high and, in the case of the second order loop, probably beyond the limit of validity of the narrow-band phase modulation assumption. For the second pair of sidebands, at a frequency separation of $\pm f_{ref}/8$ from the carrier, the levels are −25 dBc and −39 dBc, respectively, and for the third pair of sidebands, at a frequency separation of $\pm 3f_{ref}/16$ from the carrier, the levels are −32 dBc and −49 dBc, respectively. Clearly these sidebands levels are quite unacceptable and a compensation technique is essential. If the compensation were perfect then, of course, all fractional-N sidebands would be eliminated. In practice, considering the open-loop nature of the technique, the compensation waveform might reasonably be maintained to within ±1% or so of the ideal, which would be sufficient to reduce sidebands by 40 dB to a quite respectable −61 dBc if a third order loop is used.

The sideband structures obtained in this example are shown in figure 8.4, where it is clear that the third order loop offers a considerable improvement and is certainly to be preferred. Accurate and stable production of the compensation waveform is a critical area of this approach to frequency synthesis. The open-loop compensation waveform and parameters such as phase detector gain must be maintained virtually constant over a wide operating temperature range − which is not easy to achieve and is the main weakness of the technique. It is possible to use an additional control loop to fine tune the compensation gain so that thermal drifts are minimised. This can be implemented by separately detecting the amplitudes of the phase detector ripple and compensation waveform (before they are combined) and applying a low speed control loop to match these two amplitudes by adjustment of the compensation. This may produce better performance, though it still relies on careful analogue circuit design.

It should be appreciated that the sideband levels of figure 8.4 represent the worst case situation and other fractional ratios will result in wider sideband separations and lower amplitudes. In fact, consideration of diagrams such as those of figure 8.2 and 8.3 reveals that the sideband-to-carrier separation increases linearly with p to a peak of $f_{ref}/2$ when $p = q/2$

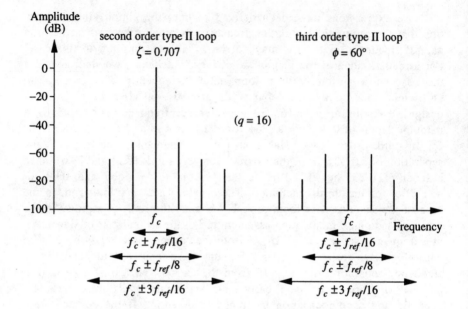

Figure 8.4   Comparison of the predicted worst case fractional-N sideband levels in second and third order loops (with 40 dB improvement due to compensation)

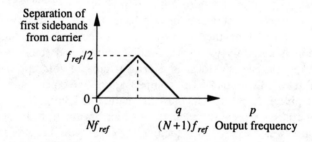

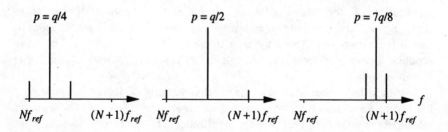

Figure 8.5   Sideband separations in a fractional-N synthesiser

and then decreases linearly to zero when $p = q$. This simple relationship is shown in figure 8.5. Evidently, the widest sideband separation, of $f_{ref}/2$, occurs when the loop division ratio is $(N + 1/2)$ and this results in the lowest sidebands since the suppression action of the loop is greater at higher modulation frequencies. It is also apparent that at least one of the first pair of sidebands will always exist at a frequency of Nf_{ref} or $(N + 1)f_{ref}$, which betrays the fact that a fractional-N method is in use.

8.2 All-digital loops

PLL circuits usually involve a combination of analogue and digital components and techniques. The analogue elements of such circuits suffer from the usual undesirable features such as reliance on component tolerances, temperature drifting, DC offsets and inflexibility in the adjustment of circuit parameters. It would clearly be attractive to implement most, or all, of the loop design in a digital fashion though, as we shall see, some parts of the loop are more amenable to digital techniques than others.

8.2.1 Frequency divider and phase detector

Taking the case of a basic frequency synthesiser, the variable ratio divider in the feedback path is the only truly digital component – based on a digital circuit operating in a digital fashion. It might be thought that a sequential phase detector falls in the same category, since it consists of digital components. In fact this is not so because such phase detectors generate a pulse train with a continuously variable duty cycle dependent on the phase difference between the two input signals and, necessarily, require output low-pass filtering in order to generate a DC signal to represent the input phase difference. A truly digital phase detector may be defined as a device which provides a *digital* output (quantised to a certain resolution) in parallel or serial format in response to the phase difference between two input signals. It must, also, operate using digital techniques alone – with no intermediate analogue stages.

An example of a truly digital phase detector is shown in figure 8.6. This circuit is the equivalent of an RS flip-flop phase detector, with an operating range of $\pm\pi$ radians, and consists of an n-bit counter connected to an n-bit latch – which is essentially a set of D-type flip-flops with a common clock. It operates by measuring the time difference, quantised to the clock period, between the rising edges of the two inputs. The counter is clocked at a rate of $f_c = 2^{n+1}f_{ref}$ so that the output derived from the most significant bit is of the reference frequency. In fact, the most significant bit must be inverted to provide a reference signal with a rising edge

coincident with a counter value of zero. The counter output is latched on
the rising edge of the loop feedback signal, V_ϕ, obtained from the VCO via
any frequency divider so that this latched value, N_{latch}, say, is proportional
to the time and hence phase delay between the reference and feedback
signals. The phase difference as a function of the latched value and
counter resolution is therefore

$$Phase\ difference\ =\ 2\pi\frac{N_{latch}}{2^n}\ \ rads\ \ where\ \ 0 \le N_{latch} \le 2^n - 1 \qquad (8.7)$$

and the smallest phase change which can be resolved is

$$Resolution\ =\ \frac{2\pi}{2^n}\ \ rads \qquad\qquad (8.8)$$

which indicates that the clock frequency must be much higher than the
reference frequency in order to obtain good resolution. From equation
(8.8), the 10-bit implementation of figure 8.6 has a resolution of 0.35°.
Although this may sound impressive it should be realised that the resolution referred to the VCO is worsened by a factor of N – to 350°, for
example, with a divider ratio of 1000. The output of the circuit varies
between 0 and $(2^n - 1)$ to represent a phase difference variation from 0 to
2π rads between the input signals. As with the analogue version of this
phase detector, the output must be offset by half the range (2^{n-1}) to allow
the loop to lock in the centre of the linear region of the phase detector
characteristic, at a phase difference of π rads.

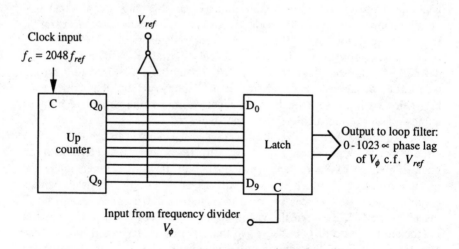

Figure 8.6 Digital equivalent of an RS flip-flop phase detector

The classic phase/frequency detector, with an operating range of $\pm 2\pi$ rads, is more attractive because of its superior capture capability, though the operation of its digital counterpart is slightly more subtle. Referring to figure 8.7, the previous design is modified by using a counter and latch which are one bit wider and by inverting the most significant bit of the counter output on alternate cycles of the feedback signal from the frequency divider. The effect of these curious modifications is apparent from consideration of a timing diagram such as that of figure 8.8. The counter now counts from 0 to $(2^{n+1} - 1)$ over *two* complete cycles of the reference signal, V_{ref}. If, from figure 8.8, we assume that initially the feedback signal, V_ϕ, lags V_{ref} by 100 clock periods (T_c) then the counter output will be alternately latched at values of 100 and $(2^n + 100)$. Now, by inverting the most significant counter bit on alternate cycles, the latched value remains constant at either 100 or $(2^n + 100)$, depending on the initial state of the toggling D-type flip-flop output, V_K. If, as in figure 8.8, we now assume that V_ϕ leads V_{ref} by 100 clock periods then the counter output will be alternately latched at values of $(2^{n+1} - 100)$ and $(2^n - 100)$. Again, the alternating inversion of the most significant counter bit acts to maintain the latched value constant at either $(2^{n+1} - 100)$ or $(2^n - 100)$. In effect, this means that the time (and hence phase) difference between rising edges of V_{ref} and V_ϕ are measured over a range of two periods – providing the 4π operating range expected of a phase/frequency detector. In the example of figure 8.8, the latch outputs are 1124, 1124, 924 and 924 representing a timing difference of 100, 100, −100 and −100 clock cycles around the mid-point value of 1024. The output of the circuit now varies between 0 and $(2^{n+1} - 1)$ to represent a phase difference variation from 0 to 4π rads between the input signals and the phase difference as a function of the latched value and counter resolution is

$$Phase\ difference\ =\ 2\pi \frac{N_{latch}}{2^n}\ rads\quad where\ 0 \le N_{latch} \le 2^{n+1} - 1 \qquad (8.9)$$

and the smallest phase change which can be resolved is, as before,

$$Resolution\ =\ \frac{2\pi}{2^n}\ rads \qquad\qquad (8.10)$$

Both of these phase detectors generate a parallel, digital indication of the input phase difference which is updated on each rising edge of the frequency divider output signal. These phase detectors may subsequently be interfaced with a digital loop filter, with attention given to the possible timing problems which may arise due to the fact that the phase detector output is not synchronised to the clock or reference frequencies.

A fundamental disadvantage of truly digital phase detectors is that their

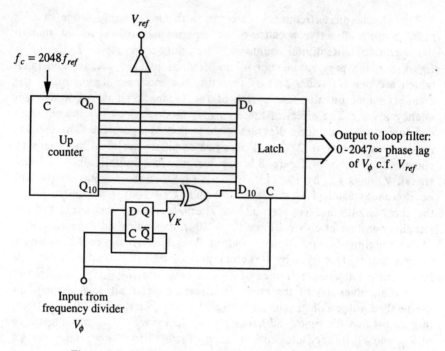

Figure 8.7 Digital equivalent of a phase/frequency detector

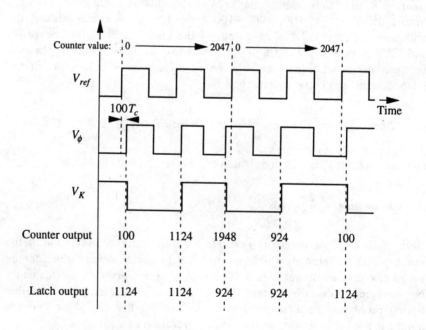

Figure 8.8 Operation of the phase/frequency detector of figure 8.7

finite resolution produces a small square-wave output component in a locked PLL which modulates the loop and results in sideband generation at a spacing of half the reference frequency. This, of course, is a result of quantisation which is only to be expected in a digital system and would not arise in its continuous analogue counterpart. It is easy to evaluate the size of these *quantisation sidebands*, as follows. The phase detector resolution is $2\pi/2^n$ rads and so in a stable, locked situation the phase detector output will contain a square-wave component equivalent to phase modulation of the reference signal by a depth of $\pm\pi/2^n$ rads and at a fundamental frequency of $f_{ref}/2$. The amplitude of the mth harmonic of this apparent input phase modulation is

$$m\text{th harmonic amplitude of quantisation phase modulation} = \frac{4}{m2^n} \text{ rads}$$

which results in VCO sideband levels (relative to the carrier) of

$$quantisation\ sideband\ levels = \frac{4}{m2^n}\left|\frac{\phi_o(m\,\omega_{ref}/2)}{\phi_i(m\,\omega_{ref}/2)}\right|\frac{1}{2} \tag{8.11}$$

Substituting the closed loop transfer function, for a second order type II PLL, and assuming that the loop natural frequency is much smaller than the reference frequency, quantisation sideband levels are given by

$$quantisation\ sideband\ levels = \frac{8N\zeta}{m^2 2^n}\left[\frac{\omega_n}{\omega_{ref}}\right] \tag{8.12}$$

and, similarly, for a third order type II loop the relative quantisation sideband levels are

$$quantisation\ sideband\ levels = \frac{8N(\tan\phi + \sec\phi)}{m^3 2^n}\left[\frac{\omega_n}{\omega_{ref}}\right]^2 \tag{8.13}$$

As an example of the magnitude of this effect, the two previous results may be used to calculate the quantisation sideband levels for a frequency synthesiser with a division ratio, N, of 1000, a damping factor of 0.707 (second order) or a phase margin of 60° (third order), a loop natural frequency of 1% of the reference frequency and a 10-bit phase detector ($n = 10$). This results in the first pair of quantisation sidebands (at a separation of $\pm f_{ref}/2$ from the carrier) having an amplitude of −25.2 dBc with a second order loop or −50.7 dBc with a third order loop and the second pair of quantisation sidebands (at a separation of $\pm f_{ref}$ from the carrier) having an amplitude of −37.2 dBc with a second order loop or −68.8 dBc with a third order loop. These results indicate barely acceptable performance and yet are based on a phase detector with a reasonably high resolution.

The maximum possible phase resolution is determined by a practical constraint − the upper frequency limit of the counter. If, for instance, the maximum clock frequency of the logic used to implement the phase detector is 51.2 MHz and the loop reference frequency is 25 kHz then a 10-bit design of the type shown in figures 8.6 and 8.7 is viable. Of course, a lower reference frequency enables higher phase resolution to be obtained. Since these phase detector designs are based on quite simple logic elements with a relatively low overall gate-count, then an ECL implementation is possible with a considerable increase in the allowable clock frequency and, hence, resolution.

8.2.2 Loop filter

Having designed a digital phase detector, it is very easy to implement a digital loop filter of whatever order or type is required by applying standard digital signal processing techniques to the relatively simple filters used in PLL circuits. It is the loop filter and phase detector interface which are most likely to benefit from a truly digital implementation by means of the elimination of DC offsets and drifts and the ability to precisely adjust circuit parameters. For instance, a digital loop filter may easily be widened at the point of switching between channel frequencies to allow rapid tuning and then gradually narrowed as phase lock approaches to reduce reference sidebands. Digital filters may be produced in either hardware or software with the hardware version likely to be faster but less flexible. Being a digital control loop, z-transforms should be used to analyse loop behaviour, although if the clock rate is much higher than the loop natural frequency then continuous methods (such as Laplace) are a good approximation. On this note, it is easy to take an analogue filter and derive the digital equivalent, as shown in the following example.

An analogue second order type II filter has the following transfer function

$$G(s) = \frac{R_2 + 1/sC}{R_1}$$

which consists of the combination of an integrating and amplifying function. The integrator may be represented in a digital system by the following describing function

$$V_{o1}(t) = V_{o1}(t-\delta t) + \frac{\delta t}{R_1 C} V_i(t)$$

where $V_{o1}(t)$ is the integrator output, $V_i(t)$ is the filter input and $\delta t = 1/f_c$ is the clock period. Similarly, the amplifier may be represented by

$$V_{o2}(t) = \frac{R_2}{R_1} V_i(t)$$

Thus the overall filter function is described by

$$V_o(t) = V_{o1}(t) + V_{o2}(t) = V_{o1}(t-\delta t) + \frac{\delta t V_i(t)}{R_1 C} + \frac{R_2}{R_1} V_i(t)$$

$$\equiv V_o(t-\delta t) + \frac{\delta t V_i(t)}{R_1 C} + \frac{R_2}{R_1}\left[V_i(t) - V_i(t-\delta t)\right] \tag{8.14}$$

A digital equivalent of the second order type II loop filter is shown in figure 8.9 and comprises an accumulator which performs the integrating function and an adder and scale factor to perform the amplifying function. It may be convenient to use the same clock frequency as for the phase detector, though this is by no means essential. It is, however, important that the filter is clocked at a significantly higher rate than the reference frequency to allow interpolation between the relatively infrequent changes in the phase detector output.

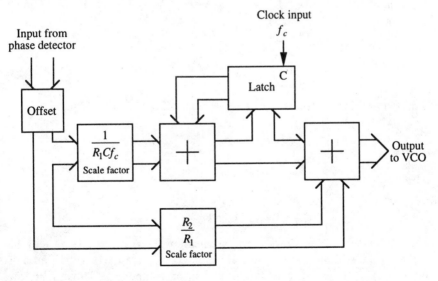

Figure 8.9 Digital implementation of a second order type II loop filter

The scale factors in figure 8.9 may be expressed in terms of the loop parameters instead of the filter component values by use of the filter design equations in chapter 3, as follows

$$\frac{1}{R_1 C f_c} \equiv \frac{N\omega_n^2}{K_p K_v f_c} \quad \text{and} \quad \frac{R_2}{R_1} \equiv \frac{2\zeta N\omega_n}{K_p K_v} \tag{8.15}$$

The resolution of the digital loop filter should be chosen to avoid degrading the resolution of the phase detector and further increasing quantisation sidebands. Because of the integrating action of the filter this means that the filter output is likely to be somewhat wider than the phase detector output.

8.2.3 VCO

The VCO is probably the most difficult element to implement digitally. In principle it is straightforward enough − requiring the generation of a sine function at a digitally controllable frequency. This may be achieved by the use of an accumulator and a sine-shaping block, as shown in figure 8.10, and, of course, requires a clock frequency of at least twice the highest

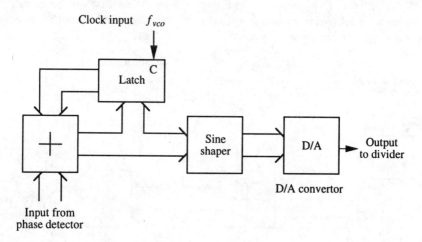

Figure 8.10 Digital implementation of a VCO

VCO frequency. Depending on the application, however, the VCO may be called on to work at microwave frequencies which are beyond the limits of digital signal processing and D/A convertor technology. Though at low frequencies it is quite feasible to use a digital VCO, at higher frequencies it is probably more sensible to resort to analogue methods and interface the loop filter to an analogue VCO by means of a high quality D/A convertor. Since the main benefits of digital implementation are realised in the filter and phase detector then there may be little loss in abandoning the digital approach in this one instance.

8.3   Computer-aided design methods

PLL design is not a trivial matter but is dependent on a number of conflicting issues usually resulting in a solution which provides an acceptable performance compromise. When embarking on a new design, a careful and probably iterative thought process is required for the design to successfully fulfil its objectives. An example of some of the issues and decisions that may be involved in the design process is given in the flow chart of figure 8.11 and the following discussion. It should be noted that this is not intended to constitute a comprehensive design method, but merely an illustration of some of the processes involved.

A PLL, of course, has four basic blocks which, although performing quite separate functions, need to successfully work together within a control loop in which the nature of any one block may affect the design or operation of the others. Perhaps a convenient starting point in a new design is the VCO. The basic parameters of this element are very easy to establish from the output frequency requirement of the circuit – which enables the VCO gain and tuning voltage range to be specified. There are, however, a number of more subtle requirements which may be of importance. These include the matter of spectral purity, which is a difficult specification to decide at the outset since the process of locking the loop, in the case of a synthesiser application, usually acts to significantly improve spectral purity and to calculate the degree of improvement the loop parameters must be known. The variation in VCO gain (i.e. linearity) may also be an important issue which will need to be addressed at a later stage in the design by means of a non-linear compensating element.

The choice of phase detector is one of the key design areas. In most cases the sequential type would be preferred because of its superior capture capabilities, but if the input S/N ratio is low or harmonic locking is required then a multiplier design would be appropriate. The phase detector gain factor is later used in the filter design, while its phase operating range may be returned to, for example, in order to determine the loop parameters required in a demodulator to avoid losing lock in the presence of a given modulation. Also, pre-phase detector filtering and/or limiting may be necessary (usually in low S/N applications) with a subsequent bearing on the loop performance.

For a synthesiser design, a frequency divider is needed, the required

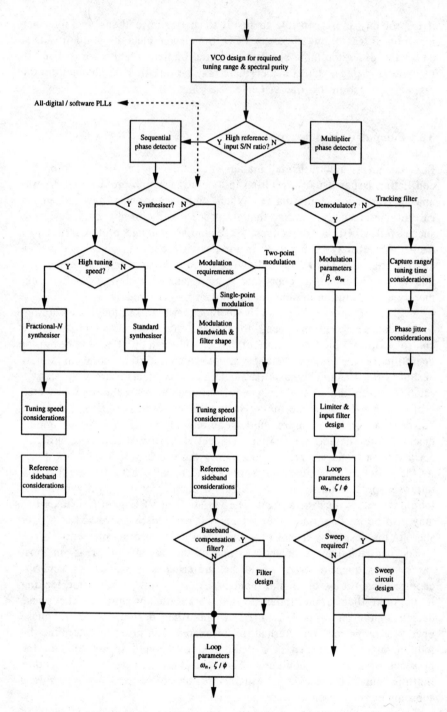

Figure 8.11 An illustration of the PLL design process

range of division ratios being obtained from the output frequency and channel spacing requirements. At low VCO frequencies it is possible to use a variable-ratio programmable divider without the requirement for a prescaler, whereas at higher frequencies a prescaler is required. If the tuning speed versus channel spacing requirement is easy to satisfy then a fixed-ratio prescaler may be adequate; otherwise dual-modulus prescaling or possibly even fractional-*N* techniques may be called upon. The divider ratio and exact technique used have a bearing on the nature and level of the reference sidebands which must be considered before finalising any design.

Finally, the loop parameters must be chosen and the loop filter designed. The filter design itself is a very straightforward matter, though the choice of parameters is often far from simple. The order and loop natural frequency are influenced by the need to suppress reference sidebands and yet allow an adequate tuning or acquisition speed. The loop natural frequency is further constrained in modulator or demodulator applications by the modulation bandwidth. If these constraints are too severe then a different modulation method, such as two-point modulation, may be appropriate. In low input S/N applications, such as receivers and data communications, the loop natural frequency is unavoidably constrained by phase jitter considerations.

The general design process lends itself to computer-aided design (CAD) methods – by means of a more comprehensive algorithm along the lines of the flow chart of figure 8.11 with a detailed library of circuit components (such as phase detectors and frequency dividers) and with a wide range of interlinked decision blocks. The majority of performance areas may, fortunately, be analysed by linear methods and are easily incorporated in a CAD package – basically by utilising (amongst others) the various analytical results derived in this book. Thus, matters such as modulation response, loop phase error, reference sideband levels and linear transient behaviour may be modelled by simple analytical methods. Capture range and tuning time, however, rely on non-linear phenomena, but may be represented by empirically-derived results or more approximate analyses. CAD methods really come into their own when it comes to modelling non-linear transient behaviour. By performing a Monte Carlo type simulation, the non-linear loop response to a transient may be built up over a succession of short time intervals to give an accurate impression of loop behaviour under conditions which may be impossible to analyse precisely. Another benefit of CAD methods is that circuit imperfections (such as phase detector DC offsets and VCO non-linearity) may easily be included to give a realistic impression of the performance of an actual circuit.

Having completed the design, the necessary components must be selected. In a few low S/N applications it may be wise to implement the design using entirely discrete components – such as a passive multiplier phase detector, op-amp loop filter and VCO built using classic RF techniques. In most applications, however, the design will be based mostly or entirely on PLL ICs of which there is a very wide choice in all technologies from CMOS to GaAs ECL. Commonly available ICs incorporate some or perhaps all of the basic PLL elements, depending on the application. In high frequency synthesisers it is usual to use an external VCO — either designed with discrete components or an MMIC device — and separate high-speed ECL prescalers in addition to the lower frequency PLL control circuitry. At lower frequencies, it is viable to use PLL ICs incorporating the VCO in addition to the other loop functions, although such VCOs are likely to have higher phase noise than their equivalent, well designed LC counterparts. It is also possible to obtain all-digital PLL ICs with their inherent advantages of device stability and design flexibility.

An example of a typical PLL IC for use with frequency synthesisers is shown in figure 8.12. This device, the MC145152, incorporates a dual D-type phase/frequency detector with a lock detector, a pair of ÷A and ÷N counters to interface with an external dual modulus prescaler and a reference oscillator and programmable divider. A complete frequency synthesiser may therefore be constructed from this device, a loop filter, VCO and an optional prescaler. The MC145152 is a CMOS device which uses look-ahead decoding to achieve a maximum operating frequency of 50 MHz. Above this frequency an additional prescaler is required. PLL ICs of this type have an unusual bias arrangement on the signal input, f_{in}, to

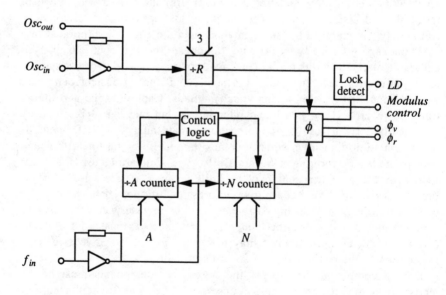

Figure 8.12 A typical PLL synthesiser IC (MC145152)

allow operation from sinewave sources of relatively low voltage (around 500 mV rms) simply by AC coupling into that input. This allows compatibility with the ECL family from which the prescaler is likely to be made. The reference frequency is divided by a choice of 8 possible values ranging from 8 to 2048 and is compared at the phase detector with the divided feedback input, f_{in}. The ÷A counter is 6 bits wide, catering for dual-modulus prescaler ratios of up to 64/65 and the ÷N counter is 10 bits wide – providing a total division ratio of as much as 65535. The divider ratios are set by parallel inputs, though there are other similar devices with serial inputs. Being a CMOS device, the phase detector pulses have an amplitude equal to the supply voltage, V_{DD}, and so the phase detector gain is simply $V_{DD}/2\pi$ V/rad.

Another classic PLL IC is the 4046 shown in figure 8.13. This contains two parallel-driven phase detectors, phase comparator 1 being an exclusive-OR gate and phase comparator 2 a phase/frequency detector. The two outputs from this latter phase detector are a tri-state indication of the phase difference, $PC2_{out}$, and a lock detect signal, PCP_{out}. Being a CMOS device, the pulse amplitude of $PC2_{out}$ is virtually equal to the supply voltage and so the phase detector gain is $V_{DD}/4\pi$ V/rad. The two phase detector inputs differ in that PCA is a self-biased input which can accept AC coupled signals of around 1 V peak-to-peak and PCB is a standard CMOS logic-level input. The 4046 also includes a VCO with a linearity of around 1% which can operate at up to 1.4 MHz. A source follower is included to buffer the VCO tuning voltage for use in modulator applications.

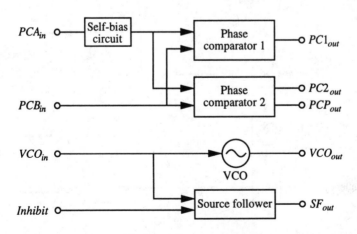

Figure 8.13 A classic PLL IC (4046)

9 Phase-Lock Applications

There are many applications of PLLs, in areas as diverse as motor speed control, digital communications and radio communications. In this chapter, a small sample of some of the more elaborate applications is presented.

9.1 Frequency translation loops

The basic loop of figure 1.1 may be modified by including a frequency translation in the path between the VCO output and the frequency divider. With this arrangement, shown in figure 9.1, the frequency at which the divider input is driven is equal to the difference between the VCO frequency, f_{out}, and the input frequency, f_{in}. Loop action adjusts the VCO frequency until this difference frequency divided by N is equal to the reference frequency. Thus the output frequency is

$$f_{out} = f_{in} \pm N f_{ref} \tag{9.1}$$

The loop may lock when the VCO frequency is either higher or lower than the input frequency, depending on the polarity of the loop feedback.

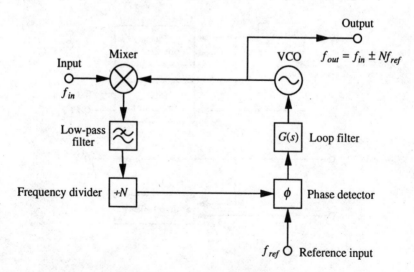

Figure 9.1 Frequency translation loop

In both cases the mixer output frequency is Nf_{ref} when the loop is locked. If, for instance, the loop is designed to lock when $f_{out} = f_{in} + Nf_{ref}$ then, referring to figure 9.2, if $f_{out} > f_{in} + Nf_{ref}$ the mixer output frequency is higher than Nf_{ref} and so the VCO is swept down until lock is obtained. Similarly, if $f_{in} - Nf_{ref} < f_{out} < f_{in} + Nf_{ref}$ then the mixer output frequency is below Nf_{ref} and so the VCO sweeps up until lock is obtained. However, if $f_{out} < f_{in} - Nf_{ref}$ then the mixer output frequency is again higher than Nf_{ref} which misleads the loop into sweeping downwards making the situation worse until the VCO latches-up at its lower limit. This latch-up problem requires that the VCO range is carefully restricted, or some other control method is used, to prevent its occurrence. Intermodulation locking is also a possible problem (see appendix G) which may require some attention.

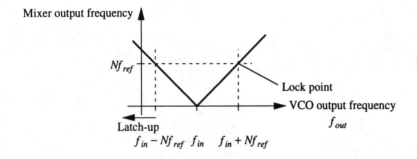

Figure 9.2 Illustration of latch-up in a frequency translation loop

Frequency translation loops have applications in frequency synthesisers, where a number of loops may be nested to obtain high resolution, and in receiver circuits involving carrier locking. It is useful to replace the standard mixer in figure 9.1 with an image-rejection mixer if the input frequency is close to the VCO frequency range in order to avoid possible image lock problems. On the other hand, if the input frequency is much smaller than the VCO frequency, then a single-sideband mixer is useful to avoid spurious loop locking and to prevent some of the undesired mixer components from propagating around the loop. Both image rejection and single-sideband mixers are described in appendix D. An example of a frequency synthesiser with application in VHF FM receivers, based on a pair of nested loops, is given in figure 9.3. It comprises an inner loop with a 1 MHz reference and a divider ratio between 87 and 106 and an outer loop with the same 1 MHz reference and a divider ratio between 117 and 126. The outer loop thus generates frequencies between 117 and 126 MHz but these are divided by a factor of 10 before being applied as an offset to the inner loop. By using a single-sideband mixer configured to produce the

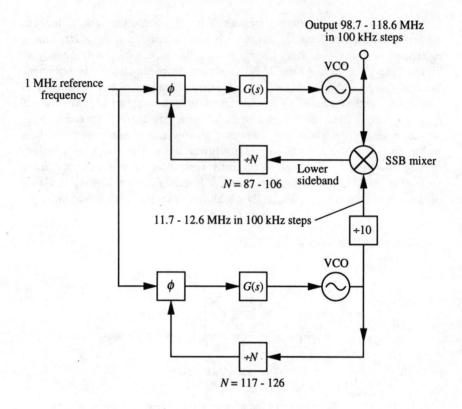

Figure 9.3   An example of a nested-loop frequency synthesiser

lower sideband, the final output frequency is the sum of that produced by
the inner loop and one tenth of that produced by the outer loop − i.e.
98.7 MHz to 118.6 MHz in 100 kHz steps. The advantage of this method
is that a synthesiser with 100 kHz resolution is obtained using loops with
1 MHz reference frequencies and so the speed of operation is ten times
that of a conventional single-loop synthesiser of the same resolution.
When mixers are used in nested loops it is important that their rejection of
unwanted mixer components is high or the loop output will suffer from
high sideband levels in a similar manner to reference sidebands. In the
case of a loop using a single-sideband mixer, it is most important that the
carrier and unwanted sideband rejection are good. Typically, the carrier
rejection might be around 40 dB and unwanted sideband rejection around
25 dB. The effect of carrier and/or unwanted sideband leakage on loop
behaviour may be analysed by reference to figure 9.4 which shows a
desired mixer component of frequency f_o and an unwanted component of
frequency f_1 and of relative amplitude A. Assuming that the loop locks

onto the dominant component of the mixer output then the presence of the additional component will result in an apparent phase modulation at a rate of $(f_1 - f_o)$ Hz and at a depth of A radians (assuming that A is quite small) around the carrier frequency f_o. This phase modulation will be modified (probably reduced) by loop action so that the phase modulation present on the VCO of the main loop will be of depth:

$$\Delta\phi_o = A\left|\frac{\phi_o(f_1 - f_o)}{\phi_i(f_1 - f_o)}\right|$$

and, from appendix B, the associated sideband levels will be exactly half of the narrow-band phase modulation index,

$$sideband\ levels = \frac{A}{2}\left|\frac{\phi_o(f_1 - f_o)}{\phi_i(f_1 - f_o)}\right| \tag{9.2}$$

It might be expected that these sidebands will always appear on the VCO spectrum at spacings of $(f_1 - f_o)$ around the carrier frequency; however, due to the sampling nature of the sequential phase/frequency detectors almost exclusively used in synthesisers, the phase modulation

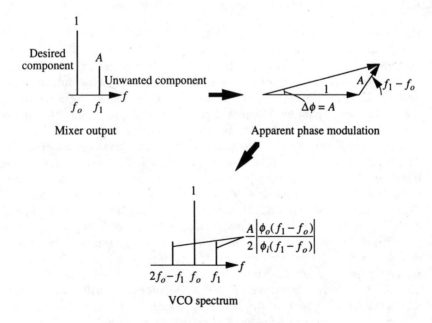

Figure 9.4 Impact of spurious mixer components on the PLL output spectrum

rate at the output of the mixer, $(f_1 - f_o)$, is actually aliased to a rate of no more than half the loop reference frequency which, for certain output frequencies in a nested synthesiser, could be as low as the channel spacing (100 kHz in the example of figure 9.3). Since the loop filter is likely to offer only limited suppression of such low rate phase modulation then it is imperative that the suppression of the SSB mixer is as good as possible in any practical design using this technique. The sidebands resulting from such nested loops are very similar to those formed in a fractional-*N* synthesiser, though the mechanism is very different. At worst, the VCO sideband levels due to this effect are equal to half the unwanted sideband suppression ratio – resulting in sidebands of an uncomfortably high level of around −30 dBc.

A completely different technique sometimes used in high resolution, high agility frequency synthesisers is direct digital frequency synthesis. This is completely different to the phase-locked approach and involves the direct computation of the digital representation of a sinewave, of the desired frequency, which is then converted to analogue form. Direct digital synthesis is capable of very high resolution and rapid switching between frequencies, but is limited to relatively low carrier frequencies (dependent on logic speeds) and tends to be costly. For these reasons, a combination of direct synthesis and nested PLLs may be used to obtain a good compromise between tuning agility, resolution and frequency coverage.

9.2   The Costas loop BPSK demodulator

A BPSK signal is phase modulated with a deviation of 0 or π rads representing the two binary states. A PLL containing a modulo-π phase detector is therefore able to lock to this signal without being affected by the modulation, enabling coherent demodulation of the signal. This is the principle of operation of the Costas loop demodulator shown in figure 9.5.

Suppose the circuit is to demodulate an input signal represented by $A(t)\cos(\omega t - \phi_i)$, where $A(t)$ is a bipolar signal, $\pm A$ say, producing the phase modulation. The signal is applied in parallel to two analogue multipliers driven by quadrature VCO signals, $\cos(\omega t - \phi_v)$ and $\sin(\omega t - \phi_v)$. Their outputs consist of a double-frequency term, which is filtered out, and a DC component proportional to $A(t)\cos(\phi_i - \phi_v)$ and $A(t)\sin(\phi_i - \phi_v)$, respectively. These signals are multiplied together so that the final phase detector output is proportional to $A^2(t)\sin2(\phi_i - \phi_v)$ – representing a modulo-π phase detector. Since $A(t)$ is either $+A$ or $-A$ then the phase detector is clearly insensitive to this modulation and will quite happily adjust the VCO to lock to the incoming signal, by forcing $\phi_i = \phi_v + n\pi$. The output of the analogue multiplier driven by the 90° leading VCO signal is therefore $\pm A(t)$ and may be level detected to produce a data

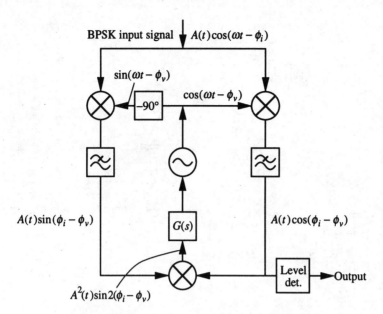

Figure 9.5 Costas loop BPSK demodulator

output. It should be noted that the ambiguity in the polarity of the output is a fundamental feature of any BPSK receiver and is resolved by means of the data coding (for instance, by use of preambles).

The Costas loop is a useful way of demodulating BPSK signals having low S/N ratios (and a more elaborate Costas loop arrangement exists to demodulate QPSK signals). Loop noise performance is, however, impaired – particularly at very low input S/N ratios – compared to loops using conventional multiplier phase detectors, due to the two-fold multiplying action of the phase detector. This can be represented as a degradation in loop S/N ratio given by, from appendix E,

$$\frac{S/N_L'}{S/N_L} = \frac{S/N_{in}}{4S/N_{in} + 1/2} \tag{9.3}$$

The deterioration in performance is therefore at least 6 dB and increases at very low input S/N ratios, although this increase is of little significance since the bit error rate would be very high under such conditions. If the input S/N ratio is good (>15 dB, say) then a standard PLL arrangement with a digital phase detector may be used as a demodulator. This is substantially simpler than the Costas loop approach and may be preferred in less demanding applications.

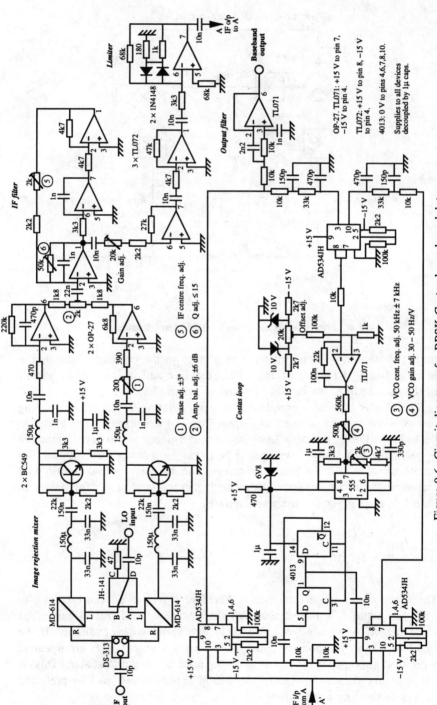

Figure 9.6 Circuit diagram of a BPSK Costas loop demodulator

An example of a Costas loop BPSK receiver is given in the circuit diagram of figure 9.6 and the associated photograph of figure 9.7. This design functions as a superheterodyne receiver sensitive to input signals of around 1.5 GHz which are down-converted by means of an image-rejection mixer to an IF frequency of 50 kHz. An op-amp band-pass filter selects the required IF signal bandwidth and the signal is then limited and passed on to a Costas loop demodulator. This demodulator comprises three phase detectors in the form of the AD534JH four-quadrant multipliers, a simple VCO based on a 555 timer and a quadrature circuit consisting, conveniently, of a ÷4 switch-tail divider. A second order type II loop is employed with the demodulated output derived from a low-pass filtered version of the leading phase detector output. Since the Costas loop is (deliberately) unresponsive to BPSK modulation on its input signal then loop parameters are dictated by more practical considerations, such as capture range, rather than by any need to accommodate the signal bandwidth. To illustrate this point, the circuit in figure 9.6 was designed to demodulate data at a rate of 4800 bits/s and yet has a loop natural frequency of a mere 300 Hz.

Figure 9.7 Photograph of the Costas loop BPSK receiver of figure 9.6

9.3   A phased array controller

A rather different application, shown in figure 9.8, makes use of the highly
linear properties of phase/frequency detectors to produce a linearly adjustable phase shifter. This is achieved by taking a standard PLL and simply
inserting a DC offset voltage, V_K, between the phase detector and loop
filter which causes the VCO phase to shift and the loop to stabilise at a
different point on the phase detector characteristic. The resulting phase
shift behaviour as a function of control voltage is a reflection of the phase
detector characteristic which, in the case of a phase/frequency detector,
gives a linear range of up to $\pm 2\pi$ radians though, of course, a range of $\pm \pi$
radians is all that is actually required. In practice, as is evident from the
experimental results of figure 9.9, the response is limited to slightly less
than this range (due to speed limitations in the logic devices) but the
circuit is certainly capable of good linearity.

An application of this technique which was investigated by Houghton
and the author (reference 7) is as a controller for a linear phased array.
Figure 9.10 shows how a cascade arrangement of identical PLL phase
shifters can produce a linearly increasing phase taper which may be used
as a series of local oscillators to control the phase on transmission (in this
example) or reception from a number of similar antenna elements arranged
in a uniformly-spaced array. The linear phase taper so produced acts to
adjust the pointing direction of the main beam formed by the linear
antenna array. The relation between beam pointing angle and inter-element
phase shift may easily be deduced by equating the phase shift through free
space in a pointing direction, θ, to the electrical phase shift introduced
between adjacent elements, ϕ:

$$\frac{2\pi d \sin\theta}{\lambda} = \phi$$

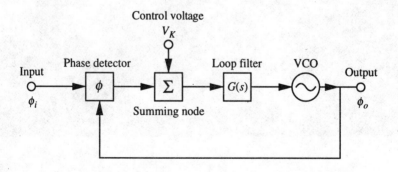

Figure 9.8   Basic arrangement of a PLL phase shifter

and so the beam pointing direction, measured with respect to the array normal, is given by

$$\theta = \sin^{-1}\left[\frac{\lambda\phi}{2\pi d}\right] \qquad (9.4)$$

The choice of loop natural frequency in this application is dependent on the speed at which the array needs to be scanned. Radar applications, for instance, are likely to be far more demanding than communications applications and are therefore likely to require much higher loop natural frequencies. The prototype system developed in this work used four PLL phase shifters operating at 41 MHz controlling four monopole elements transmitting at 1.5 GHz. The design of each PLL phase shifter is shown in figure 9.11. It comprises an LC varactor-tuned VCO with four buffered outputs, a phase/frequency detector made from two D-type flip-flops and an AND gate and a second order type II loop filter. ECL devices are used, requiring a +3 V output 'ground' reference when operated from +5 V and 0 V supply rails. The diode in the feedback path of the op-amp loop filter is required to prevent the VCO tuning voltage from becoming negative and forward-biasing the varactors. The complete phased array system consists of four of these PLL phase shifters, with four frequency convertors translating the initial 41 MHz signals up to 1.5 GHz, some DC offset trims and four half-wavelength spaced quarter-wave monopoles.

The prototype array and controller were tested in an anechoic chamber where it was found that the mean beam pointing error over a ±70° range of pointing angles was 0.9° – a quite pleasing result demonstrating the potential accuracy obtainable with the technique. A photograph of the phased array and controller prepared for measurements in an anechoic chamber is given in figure 9.12.

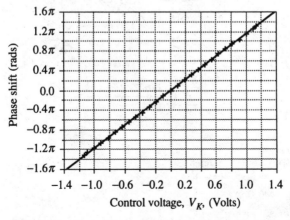

Figure 9.9 Measured response of a PLL phase shifter

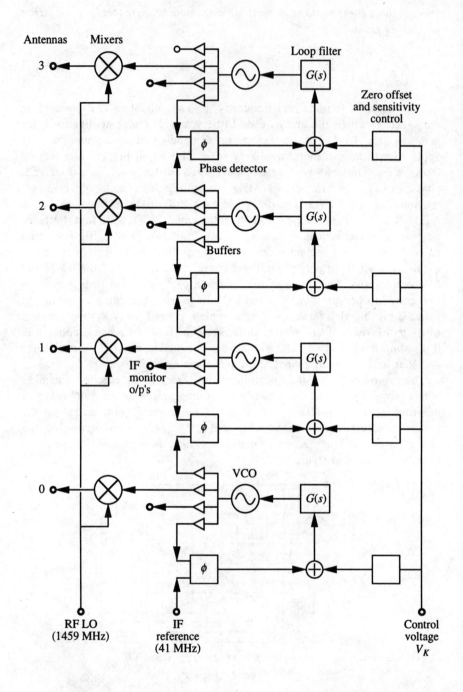

Figure 9.10 Linear phased array controller based on PLL phase shifters

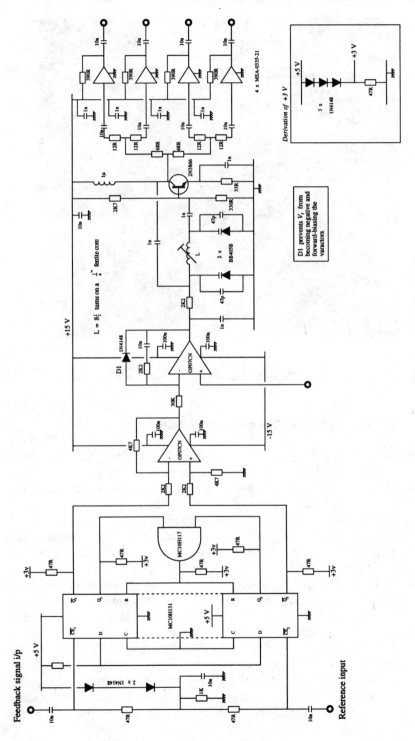

Figure 9.11 Circuit diagram of a PLL phase shifter

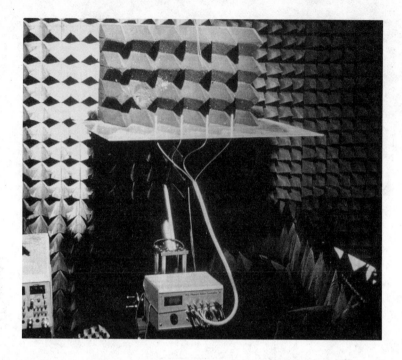

Figure 9.12 Photograph of the PLL phased array controller of figures 9.10 – 9.11

9.4 A PLL using an extended range phase detector

There is a wide variety of phase detector designs in addition to those which are featured in chapter 2. Many of these are variations on a theme although there are some that are truly different from the standard designs. An example of an unusual phase detector offering certain advantages over conventional types is considered in this section.

Figure 9.13 shows the basic design of an extended range phase detector using dual up/down counters. Each rising edge of input 1 clocks the up counter, whereas each rising edge of input 2 clocks the down counter. The digital outputs of these counters are summed in a full adder and then converted to an analogue signal by means of a D/A convertor configured to perform continuous conversion. This circuit is clearly an example of a sequential phase detector with much greater memory than the standard phase/frequency detector and, accordingly, much greater operating range. Operation may be easily understood with reference to the example input signals of figure 9.14. If, initially, the two inputs are of the same frequency but of different phase then the up counter counts up at the same rate as the down counter counts down. The output of the full adder therefore alternates between two adjacent values (8/9 in this example) with a duty cycle dependent on the phase difference between the two inputs. The mean value of the phase detector output (derived via a D/A convertor) thus indicates the phase difference between the two input signals in a very similar manner to a standard phase/frequency detector. The purpose of the differential amplifier is to provide a DC offset so that the phase detector may give a symmetrical swing around zero volts. If the frequency of input 2 now increases, then the down counter counts more rapidly than the up counter and the overall phase detector output steadily declines − reflecting the accumulated phase slippage between the two inputs. A unit change in the full adder output represents 2π radians of input phase change so that, in this example, a detector using 4-bit counters has an operating range of some $\pm15\pi$ radians. In general, a phase detector of this kind using n-bit counters has an operating range of $\pm(2^n - 1)\pi$ radians.

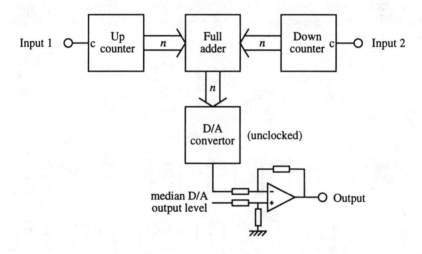

Figure 9.13 An extended range phase detector using dual up/down counters

There is no fundamental limit on the size of the counters and so it is possible to use this phase detector in highly dynamic applications − such as with rapid frequency-swept PLLs or agile synthesisers − with linear phase detector operation at all times. This means that phase and frequency transients are described by linear analytical processes and cycle slips due to dynamic effects will never occur and so phase continuity will always be preserved. However, being a sequential phase detector, it is important that S/N ratios are high, otherwise cycle slips may occur due to noise. Also, since the operating range of the phase detector is very wide, it is important that the D/A convertor and the following analogue circuitry are of low noise design in order to avoid the introduction of excessive noise into the loop.

One of the advantages of this type of phase detector is that an increase in tuning speed is possible. This can be illustrated with reference to the following frequency synthesiser example. Using the transient response equations developed in chapter 3 for a second order type II loop, if we have a frequency synthesiser where one of the phase detector input signals is subjected to a frequency step of $8\pi e\omega_n$ then the resulting phase error transient will peak at 8π radians at time $\omega_n t = 1$. This phase error is well beyond the $\pm 2\pi$ operating range of a standard phase/frequency detector and so this type of phase detector will revert to a non-linear mode of operation shortly after the start of the frequency step (at time $\omega_n t = 0.102$ when the phase error has reached 2π radians). During this non-linear phase the phase/frequency detector output voltage is approximately half of its peak value, πK_p volts, (assuming that the two input frequencies are quite similar) and the steady-state VCO ramp rate obtained with a second order type II loop is

$$\frac{d\omega_o}{dt} = \frac{K_p K_v \pi}{R_1 C} \equiv N\pi\omega_n^2 \text{ rad/s}^2 \text{ since } \omega_n = \sqrt{\frac{K_p K_v}{NR_1 C}} \qquad (9.5)$$

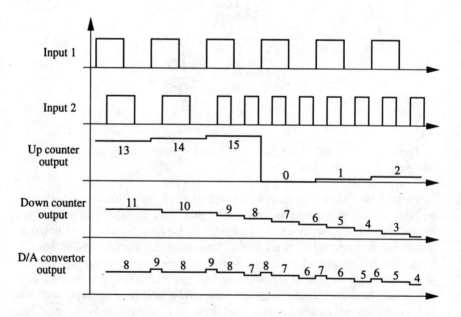

Figure 9.14 Example of operation of the phase detector of figure 9.13, with $n = 4$

The fed-back VCO frequency therefore ramps at a mean rate of $\pi\omega_n^2$ until the frequency error is within the linear range of operation, $2\pi e\omega_n$, after which the remainder of the transient is described by the usual linear process. A comparison of the performance of two PLLs with identical loop parameters but different phase detectors, based on the approximate piecewise argument given above, is shown in figure 9.15. This plot shows the significant increase in tuning time arising from the non-linear mode of operation of the phase/frequency detector in contrast to the entirely linear operation of an extended range phase detector which offers, approximately, a halving of the tuning time in this example.

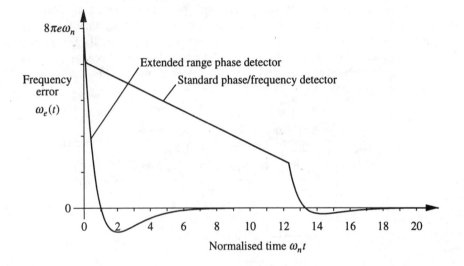

Figure 9.15 Tuning speed comparison of an extended range phase detector with a standard phase/frequency detector

9.5 Lock-in detection

Lock-in detection is a useful technique for measuring the response of a device or system to sinusoidal stimulation in the presence of large amounts of noise. Strictly speaking it is a coherent detection rather than a phase-lock technique, based on similar operating principles to analogue multiplier phase detectors.

A basic example of an experimental arrangement making use of lock-in detection is given in figure 9.16. A signal generator is used to excite a device under test with a particular frequency, ω_s, and amplitude, A. The output of this device is presented to an analogue multiplier where it is multiplied by a phase-shifted sample of the signal generator output. The multiplier output therefore contains a DC term resulting from the product of the two coherent signals originating from the signal generator. The amplitude of this DC term is proportional to the amplitude response of the device under test at frequency ω_s and a cosine function dependent on the phase shifter setting, ϕ, and the transfer phase of the device under test, as follows:

$$V_o = A|H(\omega_s)|\cos(\phi - \angle H(\omega_s)) \qquad (9.6)$$

The phase shifter setting needs to be carefully adjusted in order to maximise the sensitivity of the system, the optimum setting of $\phi = \angle H(\omega_s)$ representing the transfer phase of the device under test, enabling both amplitude and phase to be measured. In practice, however, it may not be easy to accurately or unambiguously determine $\angle H(\omega_s)$ because it relies on an optimisation process and because analogue multipliers have only a $\pm\pi/2$ rads operating range.

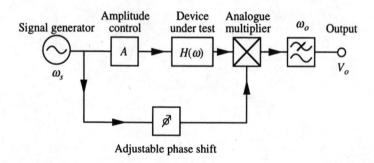

Figure 9.16 Experimental arrangement using lock-in detection

The main benefit of lock-in detection is the ease with which the effects of noise may be rejected. The analogue multiplier acts to shift any noise from the device under test by a frequency of ω_s. In so doing it translates the desired component to DC. Hence it is easy to separate noise from the desired component simply by low-pass filtering, as shown in figure 9.16. If the low-pass filter has a cut-off frequency of ω_o then only noise within the range $\omega_s \pm \omega_o$ will be admitted from the device under test. In other words, lock-in detection has the behaviour of an accurate narrow-band tracking filter – and this is where its beauty really lies.

Figure 9.17 shows an example of a circuit making use of lock-in detection. The circuit functions as a residual current device which is

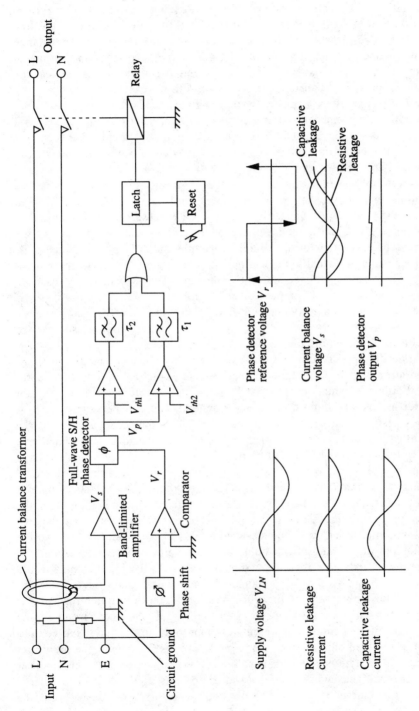

Figure 9.17 Block diagram of an RCD with immunity to capacitive leakage current

insensitive to capacitive leakage current. Residual current devices (otherwise known as earth leakage circuit breakers) provide protection from line to earth electric shocks or fire hazards by monitoring any imbalance between line and neutral currents. However, they are sometimes prone to false alarm – due partly to the steady and harmless capacitive leakage currents present in installations as a result of cable capacitances and transient suppressors – and the system of figure 9.17 addresses this issue.

Firstly, a current balance transformer and amplifier provide a sense, V_s, of any leakage current. Quite separately, a potential divider between line and earth terminals and a comparator provide a sense of the supply voltage. This is used as a reference signal, V_r, for a sample and hold phase detector which samples the sensed leakage current at just the right instants so that the circuit has maximum sensitivity to resistive leakage and minimal sensitivity to capacitive leakage. The phase shifter is required to provide a phase shift of 90° plus whatever phase shift is present in the current balance transformer. Being a sample-and-hold phase detector, its output, V_p, is a virtually ripple-free DC signal proportional to the magnitude of resistive leakage current. This signal is subsequently used to trigger the opening of a pair of relay contacts if sufficient leakage is detected. A full-wave sample-and-hold phase detector (i.e. sampling on both rising and falling edges) is specified for safety reasons so that the circuit responds within half a cycle.

A similar approach to that of figure 9.17 may allow the measurement of quadrature current in a highly inductive load – such as in industrial motor installations – which can be used to implement automatic power factor correction. An example of such an arrangement is given in figure 9.18. A series resistor acts as a current sense to detect the load current and the resultant signal is then amplified and applied to the input of an analogue multiplier. The resistor should be of low value so that only a small voltage drop is experienced or, alternatively, a current transformer may be used. A potential divider taps a small proportion of the supply voltage which is phase shifted by 90°, hard limited and applied to the other input of the analogue multiplier. Thus, if the load were purely resistive then the two analogue multiplier input signals would be in phase quadrature and so the DC output of the multiplier would have zero value. Conversely, if the load were purely inductive then the two analogue multiplier input signals would be in-phase and so the DC output of the multiplier would have maximum value. Thus, by virtue of the 90° phase shift, the multiplier response is proportional to $I_L \sin\phi$ (where $\cos\phi$ is the load power factor) and so the low-pass filtered multiplier output responds only to the inductive component of the load current, shown as I_q in figure 9.18. An analogue multiplier design of the type shown in figure 2.2 comprising an operational amplifier and an analogue switch would be ideal for this application. The filtered multiplier output is then used to control a bank of binary-weighted

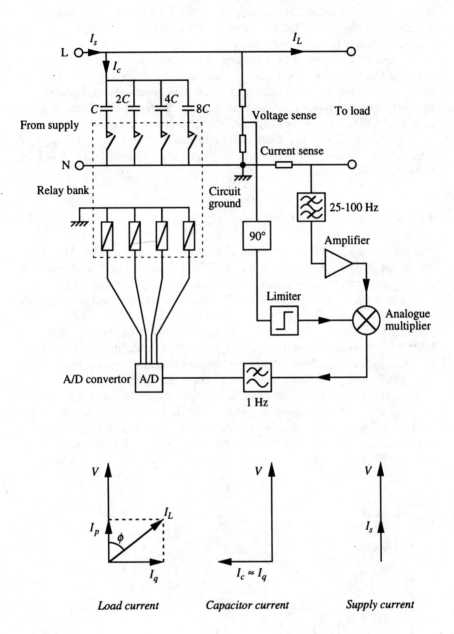

Figure 9.18 Basic arrangement of a circuit performing
automatic power factor correction

capacitors in order to cancel the quadrature component of the load current. A very low filter cut-off frequency, of around 1 Hz, is specified since it is important to reject the 100 Hz ripple on the multiplier output and because a rapid response time is unnecessary.

The circuit of figure 9.18 functions in an open-loop fashion and so it is important to use close-tolerance capacitors and to carefully calibrate the circuitry in order that the capacitor current may accurately compensate for the inductive component of the load current. Better performance may be possible with a closed-loop design, though careful attention would then need to be given to loop filter design to ensure stability, particularly in view of the relatively slow response of the relays responsible for switching in the capacitor values.

9.6   A bi-phase mark data decoder

PLLs are widely used to synchronise signals, a prime example being line and frame synchronisation in a TV set. There are also many data synchronisation applications, one example of which is featured here.

Figure 9.19 shows a bi-phase mark coded data stream. This form of data encoding is characterised by a transition at the beginning of each bit period and a further transition in the middle of a bit period to denote a logic 1. One of the features of this type of data modulation is that it has zero DC value, regardless of the data stream being transmitted, and so may be carried through AC coupled circuitry. In fact, the lower frequency limit of the data stream is $1/T$, equal to the bit rate, occurring on a long stream of logic zeros and the upper frequency limit is $2/T$ – occurring on a long stream of logic 1s. To decode the data at the receiver, a clock must be generated in synchronisation with the start of each bit period and the presence of a transition in the centre of the bit period must be determined.

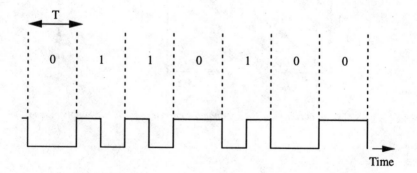

Figure 9.19   Example of a bi-phase mark coded data stream

One possible implementation of a decoder suitable for bi-phase mark data is given in figure 9.20. It consists of a simple clock recovery section based on a monostable, a PLL multiplier to generate a double-frequency clock and a simple timing/arithmetic section to detect whether there is a data transition in the middle of each bit period. Of course, there are plenty of other decoder implementations suitable for this application, some of which do not require the use of a PLL.

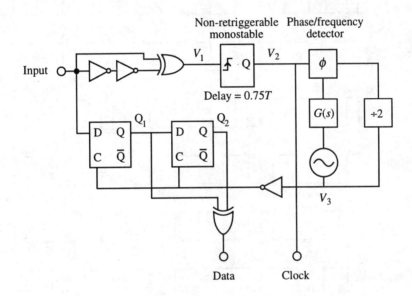

Figure 9.20 A bi-phase mark decoder

Operation may be understood with reference to the timing diagram of figure 9.21. To begin with, a pair of cascaded invertors very slightly delays the incoming data stream and an exclusive-OR gate compares the delayed and undelayed inputs to produce a positive-going spike on both the rising and falling edges, V_1. By applying this signal to a non-retriggerable monostable with a delay of between half and one times the bit period (a tolerance that most circuit designers should feel comfortable with!) a signal is produced, V_2, which has a rising edge synchronised with the start of each bit period. This is the recovered clock and is the easier part of the design. Since there may be two logic states within each bit period then a double-frequency clock is required. This is where the PLL fits in. A PLL multiplier with a ÷2 frequency divider in the feedback generates the necessary double-frequency clock, V_3. The phase detector

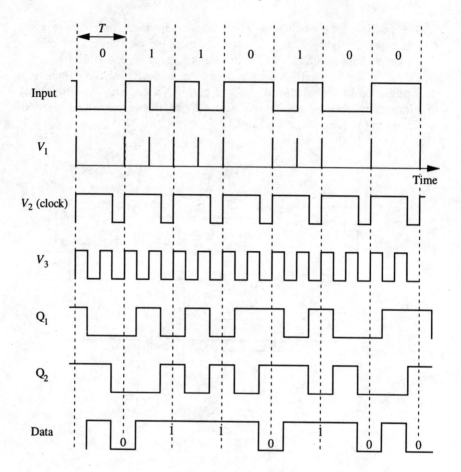

Figure 9.21 Operation of the bi-phase mark decoder of figure 9.20

must be of the sequential type because the monostable output does not
have a 50% duty cycle and, in any case, there would be no merit in using
a multiplier phase detector following a clock recovery circuit which, itself,
relies on a good input S/N ratio. In order for the double-frequency clock to
have rising edges in the centre of each *half* of each bit period, an invertor
gate is required; the resulting double-frequency clock signal then clocks
the input through a two bit shift register, the signals Q_1 and Q_2 being
replicas of the input delayed by $T/4$ and $3T/4$, respectively. By forming the
exclusive-OR of Q_1 and Q_2, the presence of a transition in the middle of
each bit period is indicated by a logic 1 pulse centred on the rising edge of
the recovered clock. Thus the output data stream contains the decoded

binary data which is synchronised with the recovered clock. The output data and clock signals enable the recovered data to be clocked into computer memory or a shift register, as required.

Concentrating on the PLL area of the design, the loop natural frequency should be around 1% of the bit rate (or less) which implies that the system may take 100 bit periods or so to stabilise. The loop filter should preferably be active, i.e. of type II, because a passive filter would result in a phase offset which could severely misalign the timing of the double-frequency clock. Many PLL ICs, such as the 4046, provide both a phase detector and VCO suitable for this purpose although the application notes invariably show a simple passive filter which, although adequate to provide capture, is somewhat lacking in many areas of performance.

Examples and Worked Solutions

Examples

1. Explain, with the aid of circuit and timing diagrams, the operation of an RS flip-flop phase detector and a dual D-type phase/frequency detector. Include sketches of the phase detector characteristics in your explanation. Derive an expression describing the characteristic of the phase/frequency detector when the input signals are of different frequencies. Also, derive a corresponding expression for the RS phase detector characteristic with input signals of different frequencies and use this to explain why loops incorporating phase/frequency detectors have superior acquisition performance.

2. Draw a block diagram and signal flow graph of a basic PLL containing a phase detector, second order type II loop filter, VCO and frequency divider. Use the signal flow graph to obtain an expression for the ratio of output phase, $\phi_o(s)$, to input phase, $\phi_i(s)$, and hence show how the loop natural frequency and damping factor are related to the loop filter component values.

A PLL frequency synthesiser with a reference frequency of 50 kHz generates an output carrier of frequency 77.3 MHz to 97.3 MHz in 50 kHz steps for use in an FM receiver. The VCO gain is 4 MHz/V in the centre of the frequency range and the phase detector gain is 1.5 V/rad. Design a second order type II loop filter with a natural frequency of 1 kHz and a damping factor of 0.707 suitable for use in this application. If the VCO gain is 5 MHz/V at 77.3 MHz and 3.5 MHz/V at 97.3 MHz then what are the loop natural frequency and damping factor at the extremes of the operating frequency range?

A/ $R_1C = 5.47 \times 10^{-4}$, $R_2C = 2.25 \times 10^{-4}$, 1180 Hz, 0.84 @ 77.3 MHz, 886 Hz, 0.626 @ 97.3 MHz.

3.   With the aid of a timing diagram, show that an exclusive-OR phase detector produces an alternating output component at twice the frequency of its input signals. For a second order type II PLL using this type of phase detector, derive an expression relating the fundamental reference sideband level on the VCO signal, due to this alternating output component, to the ratio of loop natural frequency to loop reference frequency, ω_n/ω_{ref}. You may assume that $\omega_n \ll \omega_{ref}$.

What is the fundamental reference sideband level, relative to the carrier, for such a loop with a unity divider ratio ($N = 1$), a damping factor of 0.707 and a natural frequency one hundredth of the input frequency? Comment on the level of the *harmonics* of the reference sidebands. How may (a) the phase detector and (b) the loop filter design be changed to improve the suppression of these sidebands?

A/   −43 dBc.

4.   Show, with the aid of a circuit diagram and signal flow graph, how a PLL may be (a) phase modulated and (b) frequency modulated. For a second order type II loop, derive an expression relating the frequency deviation of the loop output, $\omega_o(s)$, to the amplitude of a sinusoidal FM modulating component, $V_{fm}(s)$, and sketch the modulation characteristics for several values of damping factor.

A PLL FM modulator is required with a Butterworth response down to a 3 dB cut-off frequency of 50 Hz. What values of loop natural frequency and damping factor are required in the design? Suggest two methods of increasing the frequency range over which such a loop may be modulated.

A/   50 Hz, 0.707.

5.   Explain the meaning of the term *equivalent rectangular noise bandwidth* as applied to PLLs. Given that the noise bandwidth of a second order type II PLL is

$$B_L = \frac{\omega_n}{2}\left[\zeta + \frac{1}{4\zeta}\right] \text{ Hz}$$

derive a value for the minimum noise bandwidth and the corresponding optimum damping factor. If the phase detector gain is modified by a factor of α due to limiter signal suppression, show how this affects the noise bandwidth.

Describe what is meant by loop *phase jitter* and show that its value is given by

$$rms\ phase\ jitter\ =\ N\sqrt{\frac{B_L}{C/n_o}}\ \ rads$$

To what class of phase detector does this result generally apply?

A PLL phase demodulator with a natural frequency of 3 kHz, a damping factor of 0.8 and $N = 1$ demodulates an input signal with a signal to noise spectral density ratio of $C/n_o = 56$ dBHz. What is the rms phase jitter on the VCO signal within the loop?

A/ 9.3°.

6.   Define the terms *lock-in limit*, *pull-in limit* and *hold-in limit* as applied to a PLL. How are the values of these parameters influenced by the choice of phase detector? What, in practice, is the main effect limiting the loop capture range?

Explain, with the aid of a block diagram, a method of sweeping the VCO signal to aid the acquisition of a PLL. Derive an expression for the steady-state loop phase error as a function of the VCO sweep rate K rad/s^2. For a loop using an analogue multiplier phase detector, what is the maximum sweep rate that can possibly be supported?

A PLL, using swept VCO aided acquisition, is required to lock onto an input signal with a frequency of 10 MHz ± 50 kHz. The loop has a natural frequency of 100 Hz and uses an analogue multiplier phase detector and a VCO with a tuning range of 10 MHz ± 50 kHz. Assuming that the loop may be swept at half the maximum theoretical sweep rate derived above, what is the time taken to complete a single frequency sweep?

A/ 3.18 s.

7.   What are the advantages of using higher order loop filters? Show, by means of a circuit diagram, three methods of increasing the order of a basic second order type II loop filter? Derive expressions for the circuit components required in a third order type II loop filter as a function of the natural frequency and phase margin.

A PLL frequency synthesiser using a phase/frequency detector employs a type II loop filter with a natural frequency of 1 kHz. The loop reference (input) frequency is 25 kHz to provide 25 kHz channel spacing in the VCO output signal. Compare the reference sideband levels – due to the fundamental AC component of the phase detector output – in a second order loop having a damping factor of 0.707 with a third order loop having a phase margin of 60°.

A/ Reference sideband levels of the third order loop are 19.5 dB below those of the second order loop.

8. Describe, with the aid of a signal flow graph, the process of two point frequency modulation. Prove that this process is capable of a completely uniform frequency response regardless of the design of loop filter, deriving a value for the integration constant, A.

9. Starting with the closed loop transfer function of a third order type II PLL, derive equations for the phase and frequency transients following an input frequency step, in the case of a phase margin of 53.2°.

A frequency synthesiser is based on a VCO tuning from 150 MHz to 175 MHz, a phase/frequency detector with a $\pm 2\pi$ rads linear range, a programmable divider with division ratios adjustable between 6000 and 7000, a 25 kHz reference frequency and a third order type II loop with a natural frequency of 250 Hz and a phase margin of 53.2°. For an initial VCO frequency of 175 MHz, what is the maximum possible frequency step which will result in linear loop behaviour? What value of loop natural frequency is required to allow the loop to operate in a linear fashion during a frequency change over the complete operating range, from 175 MHz to 150 MHz? With this frequency step and loop natural frequency, how long will the loop take to tune to within 1 kHz of the final 150 MHz settling frequency?

A/ From 175 MHz to 162.825 MHz; 557 Hz; 4.44 ms.

10. Draw a block diagram and carefully labelled signal flow graph of a basic phase-locked loop comprising a phase detector, second order type II loop filter, VCO and frequency divider. Use the signal flow graph to obtain an expression for the ratio of output phase, $\phi_o(s)$, to input phase, $\phi_i(s)$, and hence show how the loop natural frequency and damping factor are related to the component values.

Explain briefly how the loop may be either frequency modulated or phase modulated, stating whether a low-pass or high-pass characteristic is obtained.

A radio transmitter uses a phase-locked loop with an output frequency of 50 MHz, locked to a 25 kHz reference. The loop is frequency modulated by a baseband signal with a lower −3dB cut-off frequency of

300 Hz. If a Butterworth characteristic is required, then what values of loop natural frequency and damping factor are needed? Taking the phase detector gain as 1.5 V/rad and the VCO gain as 2 MHz/V, design a second order type II loop filter suitable for use in this application.

$$\left[\text{You may assume the standard control equation} \quad \frac{output(s)}{input(s)} = \frac{2\zeta\omega_n s + \omega_n^2}{s^2 + 2\zeta\omega_n s + \omega_n^2} \right]$$

(University of London, 1991)

A/ 0.707, 300 Hz; $R_1 C = 3.98 \times 10^{-3}$ s and $R_2 C = 7.50 \times 10^{-4}$ s.

11. Explain, with the aid of diagrams, what is meant by the equivalent rectangular noise bandwidth of a filter. Given that the one-sided equivalent rectangular noise bandwidth of a phase-locked loop with a second order type II filter is

$$B_L = \frac{\omega_n}{2}\left[\zeta + \frac{1}{4\zeta} \right] \text{ Hz}$$

then derive a value for the minimum noise bandwidth, B_L, and the corresponding optimum loop damping factor, ζ. How much wider than this minimum value is the noise bandwidth for unity damping factor? If the phase detector gain is modified by a factor of α due to limiter signal suppression, show how this affects the loop bandwidth.

Explain what is meant by loop *phase jitter.* A PLL phase demodulator with a loop natural frequency of 3 kHz, a damping factor of 0.7 and $N = 1$, demodulates an input signal with a carrier to noise spectral density ratio of 60 dBHz. What is the rms phase jitter on the VCO signal within the loop?

(You may assume the following equation:)

$$rms\ phase\ jitter = N\sqrt{\frac{B_L}{C/n_o}} \text{ rads}$$

(University of London, 1992)

A/ 5.7° rms.

12. Figure X.1 represents the block diagram of a PLL FM demodulator. Explain briefly how the circuit operates. What values of loop natural frequency and damping factor is the circuit designed for?

Draw the signal flow graph representation of figure 1 and hence derive an expression for the transfer function $V_t(s)/\omega_i(s)$ as a function of the loop natural frequency and damping factor and the complex frequency, s. Sketch the approximate shape of this response.

The circuit is designed to demodulate an input signal with a maximum frequency deviation of 12 kHz and a modulating frequency of up to 3 kHz. Under these conditions, what is the peak phase error between the two input signals at the phase detector? Comment on the viability of the design.

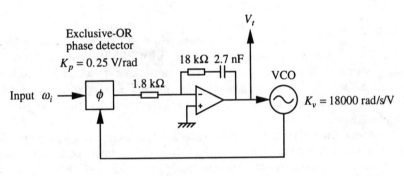

Figure X.1

(University of London, 1993)

A/ 4.8 kHz, 0.74; 79.6°.

13. Draw the block diagram and carefully labelled signal flow graph of a PLL modulator with both a phase modulation input and a frequency modulation input. Use the signal flow graph to derive expressions for the phase modulation response, $\phi_o(s)/NV_{pm}(s)$, and the frequency modulation response, $\omega_o(s)/V_{fm}(s)$. Sketch the shapes of these responses, including the roll-off rates. Explain how an additional stage may be included to achieve two-point modulation and demonstrate, mathematically, that a flat response may now be obtained.

A PLL frequency modulator uses a second order type II loop filter with a loop natural frequency of 500 Hz and a damping factor of 0.707. It is designed to be modulated over the frequency range 50 Hz to 15 kHz. What is the amplitude of the response at the lower end of the modulation range?

Suggest a method, apart from two-point modulation, that may be used to extend the frequency response.

[You may assume the closed-loop transfer function of a second order type II PLL:]

$$\frac{\phi_o(s)}{N\phi_i(s)} = \frac{2\zeta\omega_n s + \omega_n^2}{s^2 + 2\zeta\omega_n s + \omega_n^2}$$

(University of London, 1994)

A/ −40 dB.

14. Draw and carefully label the signal flow graph representation of a PLL. What is meant by the *type* and *order* of a PLL? For a second order type II loop, use the signal flow graph to derive an expression for the loop gain as a function of the loop parameters ω_n and ζ. Hence, or otherwise, derive an expression for the loop phase margin as a function of the damping factor, ζ.

A second order type II PLL is designed with a loop natural frequency of 100 kHz and a damping factor of 0.707. In order to improve performance, a first order RC filter is added immediately before the VCO tuning input. What should the 3 dB cut-off frequency of the RC filter be in order for the loop to have an overall phase margin of 55°?

[You may assume the closed-loop transfer function of a second order type II PLL:]

$$\frac{\phi_o(s)}{N\phi_i(s)} = \frac{2\zeta\omega_n s + \omega_n^2}{s^2 + 2\zeta\omega_n s + \omega_n^2}$$

(University of London, 1995)

A/ 838 kHz.

Solutions to examples

1. By comparing figures 2.10 and 2.13 it can be seen that the RS flip-flop phase detector has a similar characteristic to the phase/frequency detector for input signals of different frequency except that there is no discontinuity at $\omega_1 = \omega_2$ and the output voltage varies between V_{OL} at $\omega_1/\omega_2 = 0$ and V_{OH} at $\omega_1/\omega_2 = \infty$. This gives

$$V_p = V_{OH} + (V_{OL} - V_{OH}) \left[\frac{\omega_2}{2\omega_1} \right] \quad \text{for } \omega_1 \geq \omega_2$$

$$V_p = V_{OL} + (V_{OH} - V_{OL}) \left[\frac{\omega_1}{2\omega_2} \right] \quad \text{for } \omega_1 \leq \omega_2$$

Although RS flip-flop phase detectors have a more linear frequency detector characteristic, they are less suited for use in PLLs because the loop may become trapped in a dead zone just beyond the pull-in range of the loop where the frequency detector signal is too small to overcome DC offsets. The abrupt characteristic of phase/frequency detectors ensures that this problem does not arise and also enables faster acquisition to be achieved.

2. $K_p = 1.5$ V/rad, $K_v = 8\pi \times 10^6$ rad/s/V, $N = 1746$ at an output frequency of 87.3 MHz, $\omega_n = 2000\pi$ rad/s and $\zeta = 0.707$ in equation (3.3) gives $R_1C = 5.47 \times 10^{-4}$ and $R_2C = 2.25 \times 10^{-4}$. Taking $R_2 = 10$ kΩ gives $R_1 = 24.3$ kΩ and $C = 22.5$ nF. By noting that both ω_n and ζ are proportional to $\sqrt{(K_v/N)}$ then at the lower end of the frequency range, 77.3 MHz, these parameters are modified by a factor of $\sqrt{(5/4)}\sqrt{(87.3/77.3)} = 1.188$. Thus, at 77.3 MHz, the loop natural frequency is increased to 1180 Hz and the damping factor is increased to 0.84. Similarly, at the upper end of the frequency range, 97.3 MHz, the loop parameters are modified by a factor of $\sqrt{(3.5/4)}\sqrt{(87.3/97.3)} = 0.886$. Thus, at 97.3 MHz, the loop natural frequency is reduced to 886 Hz and the damping factor is reduced to 0.626. The variation of these parameters is quite small and is likely to have little effect on loop performance.

3. Inspection of figure 2.4 reveals that the output of an exclusive-OR gate phase detector in a *locked* PLL is a 50% duty cycle square-wave of

twice the input frequency. The amplitude of this square-wave is equal to the maximum phase detector DC output signal corresponding to a phase offset of $\pi/2$ from the locked condition – which is $(\pi/2)K_p$. By Fourier analysing the square-wave it is clear that its fundamental component (of frequency $2\omega_{ref}$) has an amplitude of $(4/\pi)(\pi/2)K_p = 2K_p$. Thus the loop may be considered to be phase modulated by this signal, its response being given by equation (4.1) which, for $\omega \gg \omega_n$, is

$$\frac{\phi_o(s)}{NV_{pm}(s)} = \frac{2\zeta\omega_n}{K_p s}$$

For a sinusoidal modulating component of amplitude $2K_p$ and frequency $2\omega_{ref}$, the peak phase deviation of the loop output is therefore

$$|\phi_o| \equiv \beta = \frac{2\zeta\omega_n N}{\omega_{ref}}$$

and this represents the phase modulation index. Recalling a basic property of narrow-band phase modulation (from appendix B), that the sideband levels relative to the carrier are one half of the modulation index, then the relative sideband levels are

$$\frac{\zeta\omega_n N}{\omega_{ref}}$$

If $N = 1$, $\zeta = 0.707$ and $\omega_n/\omega_{ref} = 0.01$ then the relative sideband level is 0.00707 or -43 dBc. In addition to this pair of fundamental sidebands, $\pm2\omega_{ref}$ from the carrier frequency, there are sidebands at the odd harmonic frequencies $\pm6\omega_{ref}$, $\pm10\omega_{ref}$ etc. Their levels, however, are very much lower because of the $1/n$ variation in the amplitude of square-wave harmonics and the first-order filtering action of the loop. Thus higher order harmonics are further suppressed by a factor of $1/n^2$: -19 dB and -28 dB for the third and fifth harmonics, respectively.

Suppression may be improved by using a phase/frequency detector (having a greatly reduced alternating component) and/or by using a higher order loop filter.

4.   Referring to equation (4.3) and figure 4.3, a damping factor of 0.707 gives a Butterworth modulation response with a 3 dB cut-off frequency of $\omega_{3dB} = \omega_n$. Thus the required loop natural frequency is 50 Hz and the

damping factor is 0.707. An extended modulation range may be obtained by using a baseband compensating filter or two-point modulation.

5. The minimum value of loop noise bandwidth may be found, in the usual way, by differentiating the expression for B_L,

$$\frac{dB_L}{d\zeta} = \frac{\omega_n}{2}\left[1 - \frac{1}{4\zeta^2}\right] = 0 \text{ when } \zeta = 1/2$$

thus the optimum damping factor is 0.5 and the corresponding minimum noise bandwidth is

$$B_L(\text{min}) = \frac{\omega_n}{2} \text{ Hz} \equiv \pi f_n \text{ Hz}$$

A loop with $f_n = 3$ kHz and $\zeta = 0.8$ has a noise bandwidth of 10.5 kHz. The input carrier to noise spectral density ratio of 56 dBHz is equivalent to 400 kHz and so the rms phase jitter is given by $\sqrt{(10.5/400)} = 0.162$ rads or $9.3°$.

6. The loop may be swept at a rate of $0.5\omega_n^2$ rad/s^2 = 197 krad/s^2 or 31.4 kHz/s. In the worst case, the VCO may need to sweep over the full 100 kHz range in which case the time taken to complete a single sweep is $100/31.4 = 3.18$ s.

7. From equation (4.1), for a second order loop the relation between VCO phase modulation, $\phi_o(\omega)$, and sinusoidal modulation, $V_{pm}(\omega)$, applied to the phase detector output is

$$\frac{K_p}{N}\left|\frac{\phi_o(\omega)}{V_{pm}(\omega)}\right| = \frac{2\zeta\omega_n}{\omega}, \text{ for } \omega \gg \omega_n.$$

Similarly, for a third order loop from equation (7.14),

$$\frac{K_p}{N}\left|\frac{\phi_o(\omega)}{V_{pm}(\omega)}\right| = \frac{\omega_n^2(\tan\phi + \sec\phi)}{\omega^2}, \text{ for } \omega \gg \omega_n.$$

Since sideband levels are directly proportional to the phase modulation index (for narrow-band PM) then the relative sideband level obtained with a third order loop compared to that obtained with a second order loop is simply

$$\left| \frac{\phi_o(\omega)_{\text{3rd order loop}}}{\phi_o(\omega)_{\text{2nd order loop}}} \right| = \frac{\omega_n(\tan\phi + \sec\phi)}{2\zeta\omega}$$

and, for $\omega = 25\omega_n$, $\zeta = 0.707$ and $\phi = 60°$ this is 0.106 or -19.5 dB. This demonstrates why third order loops are to be preferred in demanding synthesiser applications because a considerable reduction in the reference sideband level is possible.

8.   From the signal flow graph of figure 4.7, the FM characteristic obtained by applying modulation before the loop filter and via an integrator A/s is

$$\frac{\omega_o}{V_{fm1}} = \frac{K_v G(s) A/s}{1 + \dfrac{K_p K_v G(s)}{Ns}}$$

and the FM characteristic obtained by applying modulation at the usual point, immediately preceding the VCO input, is

$$\frac{\omega_o}{V_{fm2}} = \frac{K_v}{1 + \dfrac{K_p K_v G(s)}{Ns}}$$

and so the combined response is

$$\frac{\omega_o}{V_{fm}} = \frac{K_v\left[1 + \dfrac{A G(s)}{s}\right]}{1 + \dfrac{K_p K_v G(s)}{Ns}}$$

and this is equal to K_v when $A = (K_p K_v)/N$, irrespective of $G(s)$. Thus it is proved that the two point frequency modulation technique is equally applicable to any filter design – an interesting and useful result.

9.   In this example it is important to always think in terms of the frequency step experienced at the phase detector. From equation (7.21) the

maximum permissible frequency step can be found, based on a natural frequency of 250 Hz and a maximum phase change of 2π rads in a phase/frequency detector. Thus the maximum frequency step at the phase detector is $2\pi/(0.84) \times 250$ Hz = 1870 Hz. So the frequency of the feedback signal to the phase detector may rise instantaneously to 26.87 kHz when the frequency divider ratio is changed. Since the initial VCO frequency is 175 MHz then the division ratio may be reduced to as low as 175 MHz/26.87 kHz = 6513 and still stay just within the phase detector operating range. The final settling frequency with this division ratio is 6513 $\times$ 25 kHz = 162.825 MHz. This represents coverage of nearly half the full 25 MHz range.

To allow complete coverage from 175 MHz to 150 MHz with linear loop operation, the division ratio must change from 7000 to 6000 whilst the VCO is initially at the upper frequency, 175 MHz. This will result in a frequency change at the phase detector from 25 kHz to 29.17 kHz − a step of 4.17 kHz (note that the opposite frequency change, from 150 MHz to 175 MHz, results in a smaller frequency step of 150 MHz/7000 − 25 kHz = −3.57 kHz). From equation (7.21), again, the required loop natural frequency is therefore $0.84/(2\pi) \times 4.17$ kHz = 557 Hz. The output frequency error transient, for this natural frequency and for a 25 MHz frequency step, using equation (7.22), is

$$\Delta f_o(t) = N f_e(t) = 25\left[1 + 3500t - 12.25 \text{x} 10^6 t^2\right] e^{-3500t} \quad \text{MHz}$$

By using a simple iterative method, the output frequency error is found to be within 1 kHz after 4.44 ms.

It is interesting to compare the results of this question with the identical situation, though applied to a second order loop, as considered in chapter 3.

10. For a Butterworth response, a damping factor, ζ, of 0.707 is required with a loop natural frequency equal to the −3 dB cut-off frequency of 300 Hz in this case.

The loop components are calculated with N = 50 MHz/25 kHz = 2000, $K_v = 6\pi \times 10^6$ rad/s/V, $K_p = 1.5$ V/rad, $\omega_n = 600\pi$ rad/s and $\zeta = 0.707$, yielding $R_1C = 3.98 \times 10^{-3}$ s and $R_2C = 7.50 \times 10^{-4}$ s. Any sensible component values satisfying these criteria are acceptable, e.g., taking R_1 = 10 kΩ, then C = 0.398 µF and R_2 = 1.88 kΩ.

11. For minimum noise bandwidth, ζ = 0.5, resulting in $B_L = \omega_n/2$ Hz. For unity damping factor $B_L = 1.25\omega_n/2$ − which is 25% wider.

Using the first equation given in the question, a loop with f_n = 3 kHz and $\zeta = 0.7$ has a noise bandwidth of 9.96 kHz. For an input carrier to noise spectral density ratio of 60 dBHz $\equiv$ 1 MHz, using the second equation, the rms phase jitter is $\sqrt{(9.96/1000)} = 0.100$ rads or 5.7°.

12. Using the expressions

$$\omega_n = \sqrt{\frac{K_p K_v}{NR_1 C}} \quad \text{and} \quad \zeta = \frac{R_2}{2}\sqrt{\frac{K_p K_v C}{NR_1}} = \frac{\omega_n R_2 C}{2}$$

the loop natural frequency and damping factor in the given circuit are 4.8 kHz and 0.74, respectively.

Phase error needs to be evaluated under the conditions of a modulation index of $\Delta\omega/\omega_m = 12/3$ and $\omega_m/\omega_n = 3/4.8$ in the following expression:

$$\phi_e(\text{max}) = \frac{\Delta\omega}{\omega_m}\frac{(\omega_m/\omega_n)^2}{\sqrt{\left[1-(\omega_m/\omega_n)^2\right]^2 + 4\zeta^2(\omega_m/\omega_n)^2}}$$

and this gives a maximum phase error of 1.39 rads or 79.6 °

Since exclusive-OR phase detectors have an operating range of ±90° then the design will just work under the worst likely conditions and is therefore viable.

13. The FM loop response has a second order high pass characteristic with a cut-off frequency of around ω_n (depending on the damping factor). At low frequencies, the response tends to $(\omega/\omega_n)^2$ regardless of the damping factor and so at 1/10th of the loop natural frequency the response is very nearly 1% V/V or −40 dB. The frequency response may be extended by using a filter in-line with the modulation input to boost the low frequency end to complement the FM characteristic.

14. From the equation derived, the phase margin of a loop with 0.707 damping factor is 65.5°. Thus the additional filtering is responsible for 10.5° of phase shift at the frequency where the loop gain falls to unity − which is $1.554\omega_n$ for this particular damping factor (assuming that the RC filter has an amplitude response close to unity at $1.554\omega_n$). So the required RC filter has a cut-off frequency such that the phase shift is 10.5° at a frequency of 155.4 kHz and since the phase shift is simply $\tan^{-1}(\omega/\omega_o)$ then the required filter cut-off frequency is 155.4 kHz/tan(10.5°) = 838 kHz.

References

1. Gardner, F.M., *Phaselock techniques*, 2nd ed., John Wiley, New York, 1979.

2. Rohde, U.L., *Digital PLL frequency synthesisers, theory and design*, Prentice Hall, Englewood Cliffs, N.J., 1983.

3. Best, R.E., *Phase-locked loops, theory, design and applications*, 2nd ed., McGraw Hill, New York, 1993.

4. Davenport, W.B., Jnr, 'Signal-to-noise ratios in band-pass limiters', J. Appl. Phys., Vol. 24, pp. 720-727, June 1953.

5. Ascheid, G., Meyr, H., 'Cycle slips in phase-locked loops: a tutorial survey', IEEE trans., COM-30, No. 10, pp. 2228-41, Oct. 1982.

6. Brennan, P.V., 'A technique allowing rapid frequency sweeping in aided-acquisition phase-locked loops', IEE proc.-G, Vol. 138, No. 2, pp. 205-209, April 1991.

7. Houghton, A.W., Brennan, P.V., 'Phased array control using phase-locked loop phase shifters', IEE proc.-H, Vol. 139, No. 1, pp. 31-37, Feb. 1992.

Appendix A: Stability in Negative Feedback Loops

The stability of a control system consisting of a single negative feedback loop is described by two parameters of the Nyquist stability criterion: *phase margin* and *gain margin*. In order for a feedback system to be stable there must never be a frequency or frequencies where the gain around the closed-loop feedback path is greater than unity *and* simultaneously the phase shift is zero.

For a negative feedback system such as that shown in figure A.1, stability is marginal if at any frequency the loop gain, $G(s)H(s)$, has unity magnitude and a phase of 180°. It is possible to assign stability margins by inspection of the polar plot of the variation of complex loop gain, $G(s)H(s)$, over the complete frequency range from 0 to ∞.

The phase margin is defined as the difference between the loop phase $\angle[G(s)H(s)]$ and −180° at the frequency where the loop gain is unity and the gain margin is the difference in dBs between the loop gain $|G(s)H(s)|$ and 0 dB at the frequency where the loop phase is 180°. Both phase margin and gain margin must be positive in order for the system to be stable. In terms of the polar plot of $G(s)H(s)$, if the (−1, 0) point is enclosed, then the system is unstable.

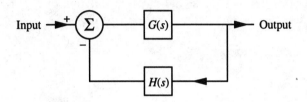

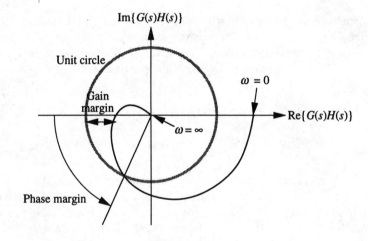

Figure A.1 Nyquist stability criteria

Appendix B: The Properties of Narrow-Band Phase Modulation

A general way of expressing sinusoidal phase modulation of a carrier by a peak value of $\Delta\phi$ and at a modulation frequency of ω_m is

$$E(t) = E_c \cos(\omega_c t + \Delta\phi \cos \omega_m t) \tag{B.1}$$

and this is identical to

$$E(t) = E_c \left[\cos(\omega_c t)\cos(\Delta\phi \cos \omega_m t) - \sin(\omega_c t)\sin(\Delta\phi \cos \omega_m t) \right]$$

In general, the frequency spectrum is described by Bessel functions in which the amplitude of the various components relates in a non-linear fashion to the modulation depth. However, if the modulation depth is small ($\Delta\phi < 0.5$ rads, say) then this equation approximates to

$$E(t) = E_c \left[\cos \omega_c t - \Delta\phi \cos(\omega_m t)\sin(\omega_c t) \right]$$
$$\equiv E_c \left[\cos \omega_c t - (\Delta\phi/2)[\sin(\omega_c + \omega_m)t + \sin(\omega_c - \omega_m)t] \right] \tag{B.2}$$

So the spectral representation of narrow-band phase modulation consists of a carrier of constant amplitude E_c and a single pair of sidebands of amplitude $E_c \Delta\phi/2$, or $\Delta\phi/2$ relative to the carrier. It is also apparent from equation (B.2) that the instantaneous phase of these sidebands relative to the carrier is $\pi/2 \pm \omega_m t$.

The phasor and spectral representations are as shown in figure B.1. The phasor diagram shows a pair of phasors rotating in opposite directions of relative amplitude $\Delta\phi/2$ and at a relative frequency of $\pm\omega_m$ around the tip of the carrier vector. The modulation phasors cross at a phase of 90° leading with respect to the carrier. If $\Delta\phi$ is small then it is easy to see that the resultant modulation will be purely phase modulation with a peak value of $\Delta\phi$. This is reflected in the spectral representation as a single pair of sidebands of amplitude $\Delta\phi/2$ corresponding to a single sinusoidal component of phase modulation.

Narrow-band phase modulation is a linear process in which a complicated modulating waveform may be reduced to its Fourier components and then converted to a power spectral plot (in dBc – decibels with respect to the carrier) merely by shifting to the carrier frequency and subtracting 6 dB, as shown in figure B.2. Conversely, this linear behaviour may be used to infer the phase modulation spectral density of a signal source known to contain only narrow-band phase modulation from its power spectrum as measured on a conventional spectrum analyser.

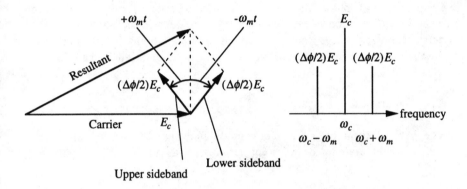

Figure B.1 Phasor and spectral representations of narrow-band phase modulation

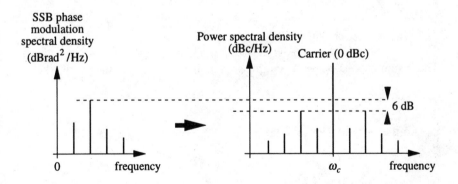

Figure B.2 Relation of narrow-band phase modulation spectrum to power spectrum

Appendix C: The Equivalence of Frequency and Phase Modulation

Frequency and phase modulation are both forms of angle modulation whereby information is carried solely by variations in the phase angle of a carrier sinusoid. This appendix shows how a simple transformation at baseband may be used to translate one form of modulation to the other.

C.1 General case

The general form of a frequency modulated signal, whereby the carrier frequency is modulated in direct proportion to an arbitrary modulating function, $V_{fm}(t)$, is

$$E_{fm}(t) = E_c \cos\left[[\omega_c + KV_{fm}(t)]t\right]$$

and similarly, the general form of a phase modulated signal where the carrier phase is modulated in direct proportion to an arbitrary modulating function, $V_{pm}(t)$, is

$$E_{pm}(t) = E_c \cos[\omega_c t + KV_{pm}(t)]$$

The instantaneous phase of the phase modulated carrier is therefore

$$\phi(t) = \omega_c t + KV_{pm}(t)$$

and so the instantaneous frequency is

$$\omega(t) = \omega_c + K\frac{dV_{pm}(t)}{dt}$$

This compares with the instantaneous frequency of a frequency modulated signal:

$$\omega(t) = \omega_c + KV_{fm}(t)$$

from which it is clear that the two forms of modulation are equivalent if the following transformation is made:

$$V_{fm}(t) = \frac{dV_{pm}(t)}{dt} \qquad (C.1)$$

This relation may be used to produce frequency modulation from a phase modulator (and vice versa) or frequency demodulation from a phase demodulator (and vice versa) simply by applying a differentiation or an integration at baseband. The four possible translations are summarised in figure C.1. Although the mathematical relation between FM and PM is precise, it should be noted that the translations derived here may not be appropriate in many practical cases. For example, by using an integrator to convert a phase modulator to a frequency modulator, the low frequency baseband components will be highly magnified and will cause large swings in the phase modulator which may be beyond its operating limits and therefore result in distortion.

C.2 Sinusoidal modulation

We shall now consider the case of sinusoidal frequency and phase modulation of peak values $\Delta\omega$ and $\Delta\phi$, respectively, at a modulating frequency of ω_m.

For FM, the instantaneous frequency is

$$\omega(t) = \omega_c + \Delta\omega \cos \omega_m(t)$$

and so the instantaneous phase is

$$\phi(t) = \omega_c t + \frac{\Delta\omega}{\omega_m} \sin \omega_m t$$

and this compares with the instantaneous phase of the PM signal

$$\phi(t) = \omega_c t + \Delta\phi \sin \omega_m t$$

thus the peak phase deviation of either signal (known as the modulation index, β) is given by the following,

$$\beta = \Delta\phi = \frac{\Delta\omega}{\omega_m} \tag{C.2}$$

This simple relation reflects the general equivalence between FM and PM in that the equivalent phase modulation may be obtained from the integral of the frequency modulation – and the integration of a sinusoidal signal component has an amplitude dependent on $1/\omega_m$.

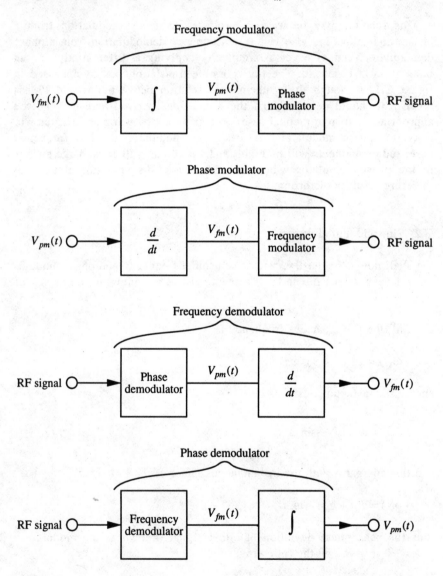

Figure C.1 Conversion between phase and frequency modulation and demodulation

Appendix D: Image-Rejection and Single-Sideband Mixers

Image-rejection and single-sideband (SSB) mixers are very useful in RF and communications applications. Image-rejection mixers behave as frequency convertors which are more sensitive to input signals lying on one side of the local oscillator than to input signals lying on the other side. This, for example, reduces the burden placed on image filters in a superheterodyne receiver design. SSB mixers are able to suppress either the sum or difference frequency of two applied sinusoidal inputs and have applications in signal generators and frequency convertor chains.

D.1 The image-rejection mixer

Figure D.1 shows the layout of an image-rejection mixer. It comprises two mixers, two quadrature (or 90°) hybrids and an in-phase splitter. Operation of the device may be understood, with some patience, by following the simple maths indicated in the figure. The outputs of the two mixers contain both sum and difference frequencies with a relative phase difference of 90° produced by the 90° hybrid in the local oscillator path. However, the relative phase of the *difference* frequency components from these two mixers changes from a 90° phase lead to a 90° phase lag around the point where the input frequencies are identical, whilst the relative phase of the *sum* frequency components remains the same. Image rejection is achieved by combining the mixer outputs in a further 90° hybrid which adds the two signals and provides two outputs – one resulting from addition with a relative phase lead of 90° and the other resulting from addition with a relative phase lag of 90°. Thus, in terms of formation of the difference frequency, one output is sensitive to input signals above the local oscillator frequency and the other is sensitive to input signals below the local oscillator frequency.

Image-rejection mixers find applications in superheterodyne radio receivers where they ease the front-end filtering requirements by inherently suppressing frequency components in the image band. In practice, suppression of the unwanted band by around 20 dB can be reliably and repeatably achieved.

D.2 The single-sideband mixer

Figure D.2 shows the layout of a single-sideband mixer. It comprises two mixers, two quadrature (or 90°) hybrids and a 180° (or sum/difference) hybrid. The arrangement is quite similar to that of an image rejection mixer. Again, the simple maths in the figure should be sufficient to explain how the mixer operates. Basically, both mixers produce components at the sum and difference frequencies, with the sum frequencies being in anti-phase and the difference frequencies being in-phase. By combining the mixer outputs in a 180° hybrid either the sum or difference frequency (i.e upper or lower sideband) may be selected.

The degree of suppression of the unwanted sideband is critically dependent on the amplitude and phase accuracy of the components within the mixer and of the circuit layout itself. The selected sideband may be represented as the vector addition of two signals of relative phase ϕ and relative amplitude A (where $\phi = 0$ and $A = 1$ in a perfect mixer) and so the amplitude of this sideband may be found from the cosine rule,

$$\text{amplitude of selected sideband} = \sqrt{1 + A^2 + 2A\cos\phi}$$

Similarly, the amplitude of the suppressed sideband may be represented as the vector addition of two signals of relative phase $(180° - \phi)$ and relative amplitude A, resulting in:

$$\text{amplitude of suppressed sideband} = \sqrt{1 + A^2 - 2A\cos\phi}$$

Thus the unwanted sideband suppression ratio is

$$\frac{\text{amplitude of selected sideband}}{\text{amplitude of suppressed sideband}} = \sqrt{\frac{1 + A^2 + 2A\cos\phi}{1 + A^2 - 2A\cos\phi}} \qquad (\text{D.1})$$

It is relatively easy to achieve a phase balance of 5° and an amplitude balance of 1 dB (i.e. $A = 0.89$) which results in a suppression ratio of 22.8 dB. This is a realistic value which can be relied on for this type of mixer. If careful attention is paid to phase and amplitude matching then performance of the order of 30 to 40 dB should be possible, though it is difficult to produce stable results very much better than these. It is worth noting that the same result (equation (D.1)) applies to the image suppression ratio obtained with an image-rejection mixer.

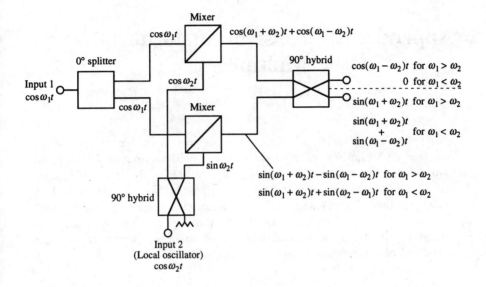

Figure D.1 An image-rejection mixer

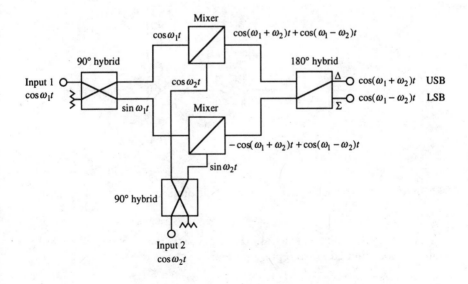

Figure D.2 A single-sideband mixer

Appendix E: Noise in Analogue Multipliers

This appendix addresses the issue of noise present on both inputs of an analogue multiplier (or mixer) and its effects on the output noise level and distribution. The situation is depicted in figure E.1. The multiplier input signals are assumed to be two carriers, of frequencies ω_1 and ω_2 and of peak amplitude E along with uniformly distributed noise of power spectral densities η_1 and η_2 V^2/Hz, respectively. Band-pass filtering is also included at the mixer inputs and is assumed to be of perfect rectangular shape and of bandwidths B_1 and B_2, respectively.

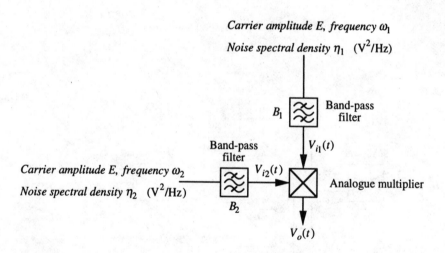

Figure E.1 Analogue multiplier operation in the presence of band-limited noise

The multiplier input signals, $V_{i1}(t)$ and $V_{i2}(t)$, may be described by a standard band-limited noise representation:

$$V_{i1}(t) = E \cos \omega_1 t + \sqrt{2 \eta_1 B_1} \cos\left[\omega_1 t + \theta_{n1}(t)\right]$$

$$V_{i2}(t) = E\cos\omega_2 t + \sqrt{2\eta_2 B_2}\,\cos\left[\omega_2 t + \theta_{n2}(t)\right] \qquad \text{(E.1)}$$

where $\theta_{n1}(t)$ and $\theta_{n2}(t)$ relate to the randomly varying noise phases and, assuming that the noises on the two multiplier inputs are uncorrelated, $\theta_{n1}(t)$ and $\theta_{n2}(t)$ are independently random variables uniformly distributed between $-\pi$ and π rads. Denoting the carrier and noise powers by units of V^2, for convenience, (or assuming a system impedance of 1 Ω) then the input carrier powers are $E^2/2$ and the noise powers are $\eta_1 B_1$ and $\eta_2 B_2$ and so the input S/N ratios are given by

$$S/N_1 = \frac{E^2}{2\eta_1 B_1}$$

$$S/N_2 = \frac{E^2}{2\eta_2 B_2} \qquad \text{(E.2)}$$

Now, the multiplier output is simply proportional to the product of the two input signals and contains terms of identical amplitudes and distributions around the sum and difference frequencies. Considering output components around the difference frequency,

$$V_o(t) \;\propto\; \frac{E^2}{2}\cos(\omega_1 - \omega_2)t + \frac{E}{2}\sqrt{2\eta_1 B_1}\,\cos\left[(\omega_1 - \omega_2)t + \theta_{n1}(t)\right]$$

$$+\frac{E}{2}\sqrt{2\eta_2 B_2}\,\cos\left[(\omega_1 - \omega_2)t + \theta_{n2}(t)\right] + \sqrt{\eta_1\eta_2 B_1 B_2}\,\cos\left[(\omega_1 - \omega_2)t + \theta_{n1}(t) - \theta_{n2}(t)\right]$$

Although the scaling is arbitrary, it is convenient to scale this result by a factor of $2/E$ to express the output noise components with respect to an output signal power of $E^2/2$:

$$V_o(t) \;\propto\; E\cos(\omega_1 - \omega_2)t + \sqrt{2\eta_1 B_1}\,\cos\left[(\omega_1 - \omega_2)t + \theta_{n1}(t)\right]$$

$$+\sqrt{2\eta_2 B_2}\,\cos\left[(\omega_1 - \omega_2)t + \theta_{n2}(t)\right] + \frac{2}{E}\sqrt{\eta_1\eta_2 B_1 B_2}\,\cos\left[(\omega_1 - \omega_2)t + \theta_{n1}(t) - \theta_{n2}(t)\right]$$

$$\text{(E.3)}$$

The four terms in this result may easily be identified as resulting from, in order, carrier $\times$ carrier, carrier $\times$ noise 1, carrier $\times$ noise 2 and noise 1 $\times$ noise 2. The output carrier and noise powers are thus,

$$Carrier\ power\ =\ \frac{E^2}{2}$$

$$Carrier \times noise\ 1\ power\ =\ \eta_1 B_1$$

$$Carrier \times noise\ 2\ power\ =\ \eta_2 B_2$$

$$Noise\ 1 \times noise\ 2\ power\ =\ \frac{2\eta_1\eta_2 B_1 B_2}{E^2}$$

As far as spectral distributions are concerned, both the carrier $\times$ noise distributions are the same as for the input, since multiplication by a carrier is simply equivalent to shifting the frequency of each noise component; however, the noise $\times$ noise distribution is rather different. Making use of the convolution theorem, whereby multiplication in the time domain is equivalent to convolution in the frequency domain, the noise $\times$ noise spectral shape is trapezoidal with a bandwidth of $B_1 + B_2$ and a flat region of width $B_1 - B_2$. Since the total power in this component is $(2\eta_1\eta_2 B_1 B_2)/E^2$ and the area within the noise $\times$ noise spectral distribution is B_1 times the peak spectral density (assuming $B_1 \geq B_2$) then the peak spectral density of this component is

$$\begin{matrix} Peak\ (noise \times noise) \\ spectral\ density \end{matrix} = \frac{2\eta_1\eta_2 B_2}{E^2} \equiv \frac{\eta_1}{S/N_2}\ \text{(from (E.2))}\ \ \text{V}^2/\text{Hz}$$

The multiplier input and output components are shown in figure E.2. The resultant noise distribution is evidently a somewhat intricate shape arising from the combination of three different noise distributions. In a typical communications application, signal 1 might be a broadband RF input signal containing a number of channels, whilst signal 2 might be a local oscillator with a small amount of relatively narrow-band noise. In such a case the IF bandwidth is likely to be less than $B_1 - B_2$ so that only a two-stage noise distribution is relevant, as shown in figure E.2.

It is interesting to derive the output S/N ratio from the mixer as a function of the two input S/N ratios, S/N_1 and S/N_2. This is easy to obtain with the aid of figure E.2:

$$S/N_{out} = \frac{E^2/2}{\eta_1 B_1 + \eta_2 B_2 + \dfrac{\eta_1 B_1}{S/N_2}}$$

$$\equiv \frac{S}{N_1 + N_2 + \dfrac{N_1 N_2}{S}}$$

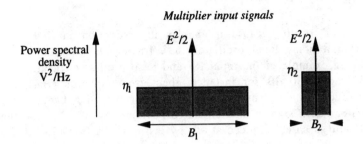

Multiplier input signals

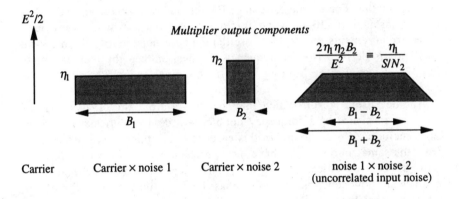

Multiplier output components

Carrier   Carrier × noise 1   Carrier × noise 2   noise 1 × noise 2
(uncorrelated input noise)

Output (signal + noise) distribution (assuming uncorrelated input noise)

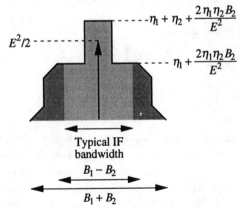

Figure E.2 Noise propagation through an analogue multiplier/mixer

$$\equiv \frac{S/N_1 S/N_2}{S/N_1 + S/N_2 + 1} \approx \frac{S/N_1 S/N_2}{S/N_1 + S/N_2} \quad \text{if } (S/N_1 + S/N_2) \gg 1 \qquad (E.4)$$

This is a useful result because it enables the reduction in S/N ratio resulting from a noisy local oscillator in a frequency convertor to be deduced. As an example, if the input RF and local oscillator signals have similar S/N ratios, 40 dB for instance, then, from equation (E.4), the output S/N ratio is reduced by 3 dB to 37 dB. On the other hand, if the local oscillator has a significantly higher S/N ratio, say 60 dB, then the output S/N ratio is barely reduced at all, by just 0.04 dB to 39.96 dB. So the conclusion here is that the local oscillator spectral purity should be significantly better then the RF signal spectral purity, which is not always an easy matter. For example, in a an FM broadcast receiver an audio S/N ratio of around 70 dB or better might be expected – which places quite severe requirements on the local oscillator phase noise purity. If this local oscillator is derived from a frequency synthesiser then the residual phase jitter may well be the limiting factor in the final audio S/N ratio of the receiver.

Another useful result which may be drawn from this analysis is the noise performance of a frequency doubler – a device with many applications including Costas loops. In this case, the multiplier inputs are derived from the same source and so have the same bandwidths, $B_1 = B_2 = B_{in}$, and noise spectral densities, $\eta_1 = \eta_2 = \eta_n$. Also, and less obvious, because the sources are now coherent, the noise phases are identical, i.e. $\theta_{n1}(t) = \theta_{n2}(t)$, which means that the two carrier × noise components are now correlated and combine in voltage rather than in power terms. From equation (E.3), the output carrier power is $E^2/2$, as before, but the output noise power is given by

$$N_{out} = \frac{(2\sqrt{2\eta_{in}B_{in}})^2}{2} + \frac{\eta_{in}B_{in}}{S/N_{in}}$$

$$\equiv 4N_{in} + \frac{N_{in}}{S/N_{in}}$$

The combination of the two carrier × noise components is evidently of twice the power that would be expected for uncorrelated inputs. The spectral distribution of the carrier × noise components is the same as before although the distribution of the noise × noise component is now rectangular, of bandwidth $2B_{in}$ and with a uniform spectral density of $\eta_{in}/(2S/N_{in})$, because of the complete correlation between the two input noise sources. This is in contrast to the case of *uncorrelated* noise of equal bandwidths on the inputs of a multiplier where the output noise × noise distribution would be triangular, of width $2B_{in}$ and with a peak spectral density of $\eta_{in}/S/N_{in}$.

The output S/N ratio is now

$$S/N_{out} = \frac{S_{in}}{4N_{in} + \dfrac{N_{in}}{S/N_{in}}} \equiv \frac{(S/N_{in})^2}{4S/N_{in} + 1} \qquad \text{(E.5)}$$

and from this result it is clear that there is at least a 6 dB degradation in additive S/N in a frequency doubler.

Where a frequency doubler is placed in the reference signal path prior to the phase detector in a PLL (such as in a Costas loop) the reduction in additive S/N is given by a modified form of the previous equation. Making the very reasonable assumption that the loop noise bandwidth is less than half the input filter bandwidth, then the reduction in loop S/N ratio is obtained by comparison of the noise spectral densities close to the carrier frequency rather than by comparison of the total noise powers. This effectively halves the contribution of the noise × noise component owing to it having twice the bandwidth of the carrier × noise components. This situation is depicted in figure E.3, from which the change in loop S/N ratio in a squaring loop is given by

$$\frac{S/N'_L}{S/N_L} = \frac{\eta_{in}}{4\eta_{in} + \dfrac{\eta_{in}}{2S/N_{in}}} \equiv \frac{S/N_{in}}{4S/N_{in} + 1/2} \qquad \text{(E.6)}$$

revealing that there is a 6 dB degradation which increases further if the input S/N ratio is very low.

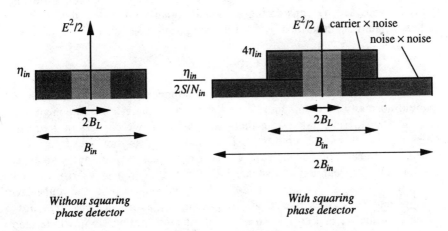

Figure E.3 PLL input noise with a squaring phase detector

Appendix F: Intermodulation

Intermodulation is a natural process which occurs when multiple signals pass through a non-linear device, system or medium, as shown in figure F.1. In many cases, the system producing intermodulation is intended to be as linear as possible (such as an amplifier), although it inevitably saturates at high signal levels and therefore has a degree of non-linear behaviour.

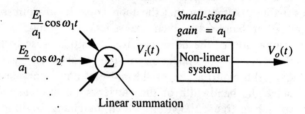

Figure F.1 The conditions leading to intermodulation

When multiple signals, each of a different frequency, pass through such a non-linear medium there is a certain amount of cross-modulation which gives rise to the generation of a large number of output components of frequencies different to (but related to) the original inputs. This is the process of intermodulation. The generation of intermodulation products may be analysed by considering the small-signal model of a general non-linear system, as described by the following Taylor series:

$$V_o(t) = a_1 V_i(t) + a_2 V_i^2(t) + a_3 V_i^3(t) + \ldots \tag{F.1}$$

In this model, a_1 represents the linear component of the system, which would normally be dominant, a_2 represents the square-law component and so on. Considering the system input to consist of two carriers of frequencies ω_1 and ω_2 and of amplitudes E_1/a_1 and E_2/a_1, as shown in figure F.1, then the linear component (a_1) gives rise to just two output signals at frequencies ω_1 and ω_2; the square-law component (a_2) gives rise to four output signals at frequencies $2\omega_1$, $2\omega_2$ and $\omega_1 \pm \omega_2$; the cube-law component (a_3) gives rise to six output signals at frequencies $3\omega_1$, $3\omega_2$, $2\omega_1 \pm \omega_2$ and $\omega_1 \pm 2\omega_2$ and so on. Each of these output signals is termed an *intermodulation product*, the *order* of which is given by the power of the non-linear term which produces the component. Concentrating on third order intermodulation (which is likely to produce the largest intermodulation products) the output signal components are

$$V_o(t) = a_3 \left[\frac{E_1}{a_1} \cos \omega_1 t + \frac{E_2}{a_1} \cos \omega_2 t \right]^3 \cdot$$

$$\equiv \frac{a_3}{a_1^3} \left[E_1^3 \cos^3 \omega_1 t + 3E_1^2 E_2 \cos^2 \omega_1 t \cos \omega_2 t + 3E_1 E_2^2 \cos \omega_1 t \cos^2 \omega_2 t + E_2^3 \cos^3 \omega_2 t \right]$$

Six third-order intermodulation products are produced, of which those at frequencies $2\omega_1 - \omega_2$ and $\omega_1 - 2\omega_2$ are closest to the original input frequencies ω_1 and ω_2. Restricting ourselves to these components, the output is therefore

$$V_o(t) = \frac{3a_3}{4a_1^3} \left[E_1^2 E_2 \cos(2\omega_1 - \omega_2)t + E_1 E_2^2 \cos(2\omega_2 - \omega_1)t \right]$$

and the levels of these two intermodulation products are

$$IM_{21} = \frac{3a_3}{4a_1^3} E_1^2 E_2$$

$$IM_{12} = \frac{3a_3}{4a_1^3} E_1 E_2^2$$

(F.2)

A similar method may be used to derive the fifth order intermodulation products, IM_{32} and IM_{23}. Third and fifth order intermodulation products are illustrated in figure F.2, where it is clear that the spacing between each of the components is equal to the difference between the input frequencies, $\omega_2 - \omega_1$. Because these intermodulation products are close, in frequency, to the input signals they may be troublesome in many applications – for example in communications links where intermodulation products are likely to appear in the centre of adjacent channels. It is also apparent, from equation (F.2), that intermodulation levels rise rapidly as the input signal levels are increased.

A commonly used method of defining the intermodulation properties of a system is the *two-tone* test where two signals of equal amplitude, E_{tone}/a_1, are applied to the input and produce symmetrical intermodulation products which, for the third order, are of amplitude

$$IM_3 = IM_{21} = IM_{12} = \frac{3a_3}{4a_1^3} E_{tone}^3$$

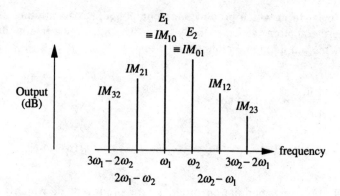

Figure F.2 Intermodulation product representation

The amplitudes of these products increase at three times the rate of the tone levels (in dB terms) and so at a certain input level the intermodulation and output tone levels might be expected to be equal. In practice, however, the system saturates and the behaviour described by equation (F.1) is no longer valid. However, a hypothetical output level may be considered where the intermodulation and tone levels would be equal in the absence of saturation – and this is termed the *intercept point* (figure F.3). The third order intercept point, IP_3, is easily found by equating the intermodulation product levels, IM_3, to the tone levels, E_{tone}:

$$\frac{3a_3 E_{tone}^3}{4a_1^3} = E_{tone} = IP_3 \quad \text{i.e.} \quad \frac{a_3}{a_1^3} = \frac{4}{3IP_3^2}$$

and this enables third order intermodulation product levels to be expressed in terms of the intercept point,

$$IM_{21} = \frac{E_1^2 E_2}{IP_3^2}$$

$$IM_{12} = \frac{E_1 E_2^2}{IP_3^2}$$

(F.3)

which may be written in dB form,

$$IM_{21} = 2E_1 + E_2 - 2IP_3$$

$$IM_{12} = E_1 + 2E_2 - 2IP_3$$

These results may easily be extended to apply to any arbitrary intermodulation component,

$$IM_{mn} = mE_1 + nE_2 - (m+n-1)IP_{(m+n)} \tag{F.4}$$

Figure F.3 shows the result of a typical two-tone intermodulation measurement of an amplifier. The device clearly has a gain of 20 dB with the output saturating at around 14 dBm. The 1 dB compression point represents the output level when the gain has dropped by 1 dB from its small signal value – which appears to be 10.5 dBm in this example. However, because two simultaneous tones are applied in this measurement, the 1 dB compression point to a single tone is actually 3 dB higher – giving 13.5 dB. Third order intermodulation clearly increases three times as rapidly as the tone levels before beginning to saturate at around 3 dBm. The linear extrapolation of these two curves crosses at 24 dBm, which is the third order intercept point and may be used to predict intermodulation levels with the aid of equation (F.4). As a general rule of thumb, the third order intercept point is around 10–15 dB higher than the 1 dB compression point.

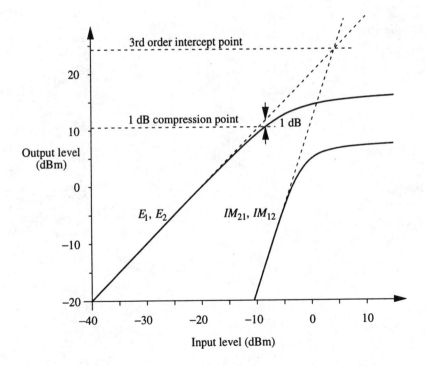

Figure F.3 A typical two-tone intermodulation measurement result

Appendix G: Laplace Transforms

The following is a summary of the Laplace transform results used in this book.

Function	$f(t)$	$F(s)$
Unit step	1	$\dfrac{1}{s}$
Unit ramp	t	$\dfrac{1}{s^2}$
Unit parabola	t^2	$\dfrac{2}{s^3}$
Exponential	$e^{-\omega_o t}$	$\dfrac{1}{s+\omega_o}$
	$te^{-\omega_o t}$	$\dfrac{1}{(s+\omega_o)^2}$
	$t^2 e^{-\omega_o t}$	$\dfrac{2}{(s+\omega_o)^3}$
Derivative	$\dfrac{df(t)}{dt}$	$sF(s)-f(0_+)$
Integral	$\displaystyle\int f(t)\ dt$	$\dfrac{1}{s}F(s)$
Final value theorem	$\displaystyle\lim_{t\to\infty} f(t)$	$\displaystyle\lim_{s\to 0} sF(s)$

Glossary

A brief definition of some of the terms used in this book:

Acquisition: The process of attaining a stable, locked state in an initially unlocked or perturbed PLL.

Closed-loop transfer function: The function representing, in complex frequency terms, the relation of PLL output phase to input phase.

Cycle slip: An abrupt and irreversible increase in phase error occurring when the instantaneous loop phase error exceeds the phase detector operating range. The magnitude of each cycle slip depends on the phase detector characteristic.

Damping factor: A characteristic of second order systems determining the shape of the frequency response to external sinusoidal excitation. In the frequency domain, a low damping factor gives rise to a pronounced peak close to the natural frequency. In the time domain, the damping factor has a significant bearing on the transient response – unity damping factor providing the minimum settling time.

Fractional-N synthesis: A technique in which rapid changes in the programmable divider ratio within a PLL produce the effect of a fractional division ratio. The technique provides sub-reference frequency resolution and overcomes the classical compromise between frequency resolution and tuning speed.

Hold-in range: The frequency range over which a loop is capable of operating, though not necessarily acquiring lock.

Lock-in range: The range of frequency offsets within which an initially unlocked loop may acquire lock without any cycle slips.

Multiplier phase detector: A device which forms the linear product of two alternating signals from which an indication of the phase difference may be obtained.

Natural frequency: A characteristic of second order systems related to, but not quite equal to, the cut-off frequency of the system response when excited by an external sinusoid. In PLL terms, the natural frequency along with the damping factor completely define the nature of the closed-loop response.

Noise bandwidth: The effective bandwidth of any real filter when presented with an input consisting of uniformly distributed noise. Otherwise referred to as *equivalent rectangular* noise bandwidth.

Order: A control definition equivalent to the mathematical order of the denominator of the closed-loop transfer function.

Phase/frequency error: The difference, in either phase or frequency, between the two signals present at the phase detector. May be a result of static conditions – such as a DC offset giving a constant phase error, or dynamic conditions – such as a frequency demodulator having time-varying phase and frequency errors as a result of modulation of the input signal.

Phase jitter: Noise-like phase variations in the VCO signal due, primarily, to additive noise at the loop input. Easily enumerated by knowledge of the loop noise bandwidth and input carrier to noise power spectral density ratio.

Phase-plane representation: A method of presenting dynamic or transient behaviour in which the loop frequency error is plotted against phase error. Of particular use in representing non-linear loop performance, such as capture in a loop containing an analogue multiplier phase detector.

Prescaler: The first in a cascade arrangement of frequency dividers, used to extend the operating frequency of a lower frequency divider. Usually of ECL technology and capable of operation well into microwave frequencies. Prescalers may be of fixed ratio or dual/triple modulus, the latter allowing integer steps in the division ratio.

Pull-in range: The range of frequency offsets within which an initially unlocked loop is capable of acquiring lock, although with a number of cycle slips.

Quantisation sidebands: Sidebands on the VCO signal arising from the quantised nature of part- or all-digital PLLs.

Reference sidebands: Sidebands appearing on the VCO signal as a result of modulation within the loop by the AC component of the phase detector output. The separation of these sidebands from the carrier frequency is dependent on the phase detector design and their amplitudes are dependent on both phase detector and loop filter designs.

Sequential phase detector: A device which responds to the relative timing between the edges of two alternating signals from which the phase difference may be obtained.

Two-point modulation: A technique in which both phase and frequency modulation characteristics of a PLL are combined to achieve either phase or frequency modulation with a completely uniform response. Requires the use of a differentiator or integrator and a carefully chosen relative weighting. The technique applies to both modulation and demodulation and is independent of the loop filter characteristic.

Type: A control definition equivalent to the number of perfect integrators present in the loop. PLLs are at least type I because of the integrating effect of the frequency to phase conversion between the VCO and phase detector. High gain active loops (i.e. containing an integrating filter) are considered type II. In general, higher loop types have superior tracking and dynamic performance.

Index